PRAISE FOR *The Dangerous Joy of Dr. Sex and Other True Stories*

"*The Dangerous Joy of Dr. Sex and Other True Stories* is a dangerous joy of literary pleasure—a compelling, spellbinding reading experience. In this book, Pagan Kennedy writes with clarity, honesty and impeccable grace."

—Lee Gutkind, editor of *Creative Nonfiction* and author of *Almost Human: Making Robots Think*

"The most daring writer I know, Pagan Kennedy prowls the shadowy, creepy, eye-popping limits of the culture where other writers fear to tread. And the stories she brings back are so tomorrow, if not so next year, that I sometimes wonder how life can ever catch up. This book is a crystal ball. Take a look: Your future is inside."

—John Sedgwick, author of *The Dark House* and *The Education of Mrs. Bemis*

PRAISE FOR *Black Livingstone,* a *New York Times* Notable Book and winner of the Massachusetts Book Award Honor in Nonfiction.

"Kennedy chronicles Sheppard's life with a near pitch-perfect combination of sympathy, drama and historical comprehension."

—*LA Times*

"Kennedy offers a smoothly written tale of Sheppard's life and is to be commended for bringing his extraordinary story to greater prominence.""

—*The Washington Post*

"The author's enthusiasm for her subject is infectious and makes the book an engaging, quick read."

—*The Houston Chronicle*

PRAISE FOR *Confessions of a Memory Eater:*

"In her absorbing and timely novel *Confessions of a Memory Eater,* Pagan Kennedy explores love, addiction, and memory in the pharmaceutical age....In fewer than 200 pages, *Confessions* packs an allegorical wallop."

—*Entertainment Weekly*
A 'Must Read for Summer '06' Full Page Feature Review

"Pagan Kennedy charges into the future with impressive dexterity....*Confessions of a Memory Eater* is just a moving portrait of mankind's chronic and untreatable case of folly."

—*New York Times Book Review*

"What Kennedy has written is a provocative novel, fast-paced and delicious."

—*Newsday*

"*Confessions of a Memory Eater,* by Pagan Kennedy is the perfect distillation of a midlife crisis, except it's a hundred times more fun and entertaining than that sounds. A former academic wunderkind, Win Duncan discovers a drug that allows you to relive any memory as if it were happening right now. But the drug, mem, works better if you take it with someone else who shares the same memory. Nostalgia becomes a consumer item for a cast of characters who are less than the sum of their pasts. Give this book to your friends and they'll be way more interesting to talk to afterward."

—*San Francisco Bay Guardian*

"Pagan Kennedy, a '90s 'zinequeen turned novelist, ventures into a surreal genre of time travel, addiction, and midlife crises in her new novel *Confessions of a Memory Eater* [which] succeeds in being both a quick, suspenseful read and a more thoughtful probe into what exactly we fear we lose with age....*Confessions of a Memory Eater* nudges readers to explore their own troves."

—*Boldtype.com*

"Telling, wistful and heartbreaking...Kennedy's narrative packs such a wrenching emotional punch."

—*Los Angeles Times Book Review*

"Kennedy turns in a surprisingly bittersweet novel with science fiction overtones and a delicious premise....[Her] easy style masks a fierce intelligence and painstaking artistry in this melancholy midlife crisis-with-a-twist. "

—*Publishers Weekly*

"I've been dreaming of the pill that would take me back to Halloween '94, when I made out with the hottest guy in the universe. Pagan's novel takes that idea—a pharmaceutical called Mem— and runs with it. The story is a total mind melt-down that kept me reading even though I was distracted thinking about that amazing night..."

—*JANE,* June/July '06

"Eschewing TV's manipulative cheap tricks, Kennedy invents a drug that allows people to access memories and refeel them, in the process raising larger questions about identity. Who (or perhaps when) is your true self? Is the realest you the one who lives in the present moment? Or is it the person you once were, at eight, at 27?...Think of it as a kind of future non-fiction— it's fiction now, but maybe not for long. In the meantime, she dramatizes the elements of time, memory, and identity in all their fluid dynamics."

—*The Boston Phoenix*

"The memory-enhancing pill at the center of Pagan Kennedy's new novel works like Proust's madeleine in overdrive: It doesn't just pave the way to relaxed reverie but allows people to relive, in detail, any past experience that they choose. That's an enticing concept for nostalgia junkies, but as *Confessions of a Memory Eater* reveals, the rewards of such a drug are inevitably trumped by its perils....With a mixture of wit and emotional honesty, fizzy prose and Technicolor descriptive passages, the premise of *Confessions* may be pulpy, but it makes a memorable point: The pleasures of youth should have an expiration date."

—*Time Out New York*

"Kennedy resists the temptation to inflict a politically correct paradigm on her story and simply lets it tell itself....The result is a page-turner that illuminates as it breaks the heart."

—*Philadelphia Inquirer*

Also by Pagan Kennedy

*The First Man-Made Man: The Story of Two Sex Changes,
One Love Affair and a Twentieth Century Medical Revolution*

Confessions of a Memory Eater

Black Livingstone: A True Tale of Adventure in 19th-Century Congo

The Exes

Spinsters

'Zine

Stripping

Platforms

The Dangerous Joy of Dr. Sex and Other True Stories

Pagan Kennedy

Santa Fe Writers Project, SFWP and colophon are trademarks.

Library of Congress Cataloging-in-Publication Data

Kennedy, Pagan, 1962-
The dangerous joy of Dr. Sex and other true stories/Pagan Kennedy.
    p. cm.
ISBN 0-9776799-3-4 (trade paper)
       1. Eccentrics and eccentricities—United States—Biography.
       2. Eccentrics and eccentricities—United States—Anecdotes.
       3. Eccentrics and eccentricities—United States—Humor. I. Title.
PS3561.E4269D36 2008
818'.54—dc22

                                                                    2008005457

Cover design by Bill Douglas at The Bang
Printed and bound in Canada

Visit SFWP's website: www.sfwp.com and literary journal: www.sfwp.org

Santa Fe • Washington, DC

TRANS 10 9 8 7 6 5 4 3 2 1

For Kevin

INTRODUCTION

I learned about secret doorways when I was a kid. One summer afternoon, my grandmother and I perched on her bed together; she balanced a book on her lap and read to me about Alice, who passed through a mirror as easily as you'd push through a curtain.

"I'm going to do that," I told my grandmother, touching the illustration, which showed a girl in striped stockings emerging into Looking Glass Land.

"Oh no, Honey. She didn't really. It was just a dream she had by the fire," my grandmother told me in her breathy Virginia accent, which turned the word "fire" into "far," and made everything sound like a question.

I knew she was wrong. Books didn't lie. Why would an author sit writing at his desk for years just to describe a world that didn't exist? And besides, this business about Alice was clearly meant to be a set of instructions. You only had to follow them to get into Wonderland. Any idiot could see that. "I'm going to do it," I said, and jumped off the bed, padding over to the other end of the room. I stood in front of my grandparent's oval mirror, which I had already decided must be magic because of its extreme age, tarnished glass, and gilt frame. I leaned so close that my breath fogged up the surface. I touched the mirror, gingerly, probing for a chink. When that didn't work, I studied the images reflected on the glass for clues: the backwards room, a curtain curled around buttery sunlight, and my grandmother propped up on the bed, her pantyhose-covered legs a curious and unconvincing shade of skin tone. Objects carried a significance that they did not have in real life. Outside the mirror-window, in the wrong-way yard, a cedar tree pointed like a gnomon with its purple shadow. What lay beyond? Was there a reverse me hiding in the topsy-turvy streets? Like Alice, I was convinced that on the other side, just out of the frame of the mirror, reality warped and buckled.

I scooped up the book, and then carried it over to the mirror, holding it open to its own image. There I saw its code, the backwards writing like a dispatch from another universe. I was on the right track. The secret had something to do with books and language, and the places where the world turns weird. But I still couldn't find the way through.

***

When I was in my 20s, a pattern of leaves on the sidewalk or the crook of a lamppost would fill me with longing. Out of nowhere, I would be seized with homesickness for a place that never existed.

When that happened, I would rush back to my desk and begin a short story. By making a spell out of words, an incantation of characters, I hoped I could travel into that other land, the place where I belonged.

In those days, I also wrote nonfiction, but only to pay the bills: I churned out last-minute blurbs for *Interview,* think pieces for *The Village Voice,* ruminations about literature for *The Nation.* In the late 1980s and early 1990s, the alternative weeklies ruled the publishing universe; they paid well enough that I could scrape together a living from them. Still, I considered myself a fiction writer. Making stuff up would be my "real" work. I needed to find the place that—more than America—I knew to be my country. But I only seemed to travel to my homeland in the saccade of reading, that instant of the eye catching up a bundle of words and transmitting a pulse of understanding in the brain. In the next moment I lost my home. Blink: I had it. Blink: it was gone. How to live there?

More so even than novels, I loved the New Journalism of the 1960s, those bad-boy books that turned reportage into a grand, show-offy art. *Fear and Loathing in Las Vegas* and *Hell's Angels* by Hunter Thompson, *The Electric Kool-Aid Acid Test* by Tom Wolfe, and anything at all by Truman Capote—these books appeared as if they had simply winked into existence, too grand to be the work of one human being. I used to stare at pages of *Electric Kool-Aid*, trying to understand how Wolfe had managed to turn raw experience into a fable. Had he used a tape recorder all through the months he spent with the Merry Pranksters? Would a machine even be able to capture conversation on a noisy bus? And once he had thousands of pages of notes, what alchemy had he used to make his book feel like an acid trip?

So for years, I was a fiction writer with a side-fetish for truth stories. Or rather, I was too intimidated by the prospect of uniting the techniques of journalism and fiction to attempt it myself. And then one day in 1996, I stumbled across an untold true story, one that I knew would make a magnificent book. Once I began that story, questions about research and technique hardly mattered. Obsession became my teacher. While reading a history of Africa, I had come across a fascinating anecdote: In the late 19th century, two American missionaries set out to explore the Congo and establish an outpost there. One of the men was black and the other white. In the United States, Jim Crow laws would have barred the two men from so much as eating together. But in the Congo, the two missionaries slept in a tent together, nursed each other through illnesses, and eventually founded a town together. After Sam Lapsley—the white man—died, William Sheppard became the first explorer ever to enter the "forbidden kingdom" of the Kuba people; he also collected evidence of atrocities that put him at the center of an international human-rights effort. The story haunted me. I could not stop wondering about William Sheppard.

I felt utterly unqualified to tell this story. Still, what choice did I have? Here was the doorway into another world, the kind I'd been looking for all my life. Sheppard seemed to hold the key. I had to understand who this exceptional man had been, how he'd managed to reinvent himself from dirt-poor preacher to internationally known explorer. I pored over his writings in an attempt to get closer to Sheppard. He came to me in dreams, in his high Victorian collar, with that glint in his eye, but he would tell me nothing, not even in the realm of sleep.

At this point, my grandma—the one who introduced me to Alice and mirrors—was dying. She'd lost most of her hearing and was completely blind, shut into her own memories. To communicate with her, you had to scream into her good ear. I wrote a letter to her about my Sheppard project; someone had to shout the message to her. And then she dictated a message to me: she had something important to tell me about Congo missionaries. Next time I traveled down South to visit her, she propped herself up on her bed, her blue eyes staring out at nothing, her hair perfectly coifed into a meringue, and told me that her first cousin had been a missionary in the Congo in the early 1900s. In that moment, I discovered another door leading from the real to the fantastic—and once again, my grandmother had opened it for me. In my researches, I had frequently come across references to Robert Dabney Bedinger; he had written the first-ever biography of William Sheppard, who had been his colleague in the Congo. Bedinger had been Sheppard's greatest defender. I had never suspected a connection to my own family. Now I knew it was in my blood. My grandmother's maiden name—signifier of her lost girl-hood identity, of her lost Alice-hood—was Bedinger.

It was at that point that the Sheppard project had evolved from hobby to avocation. I wrote up a book proposal, working to master the unfamiliar form of the "nonfiction novel." Viking Press bought the idea, and the book became my full-time job. The production of this biography felt entirely different than anything I'd done before. As a novelist and short-story writer, I'd had to force myself to park myself at my desk. To invent imaginary characters, you have to create a vast fake film set in your brain where your fake people can pace, gesture, and speak. This can be horrible, especially when your jerry-built world reveals itself to be nothing but cardboard. Now, however, I found myself working for twelve hours with no effort, pulled by my research into what had become a labyrinth of clues about William Sheppard.

The facts infected me like tropical fever, and I couldn't stop digging for more, another document, another revelation. I wrote with a new kind of vigor; I wanted everyone to see the shining shards I had dug up, to understand them and to *care*. It saddened me that so few Americans knew what had happened during the worst years of the Congo holocaust. I wrote that book under the shadow of the Rwanda genocide of 1994, disgusted that my country had done so little to intervene, and asking myself painful questions about why I failed to live up to my own values. As I pieced together William Sheppard's story, I sometimes felt I had plummeted straight through a wormhole, into another person's mind; at other times, William Sheppard seemed to run far ahead of me, impossibly remote, and the best I could do was study a boot print he'd left behind. Even when I was lost, I felt passionately in love with Sheppard and his story. Now I knew that I'd always been a nonfiction writer. And indeed, though I have returned to fiction occasionally, the true story remains my true love.

After the book came out in 2002, it won awards. Magazine editors began approaching me, presenting me with offers that I had never dared to hope would come my way. Now I suddenly could fly around the country and chase down true stories as they unfolded. Of course, I'd been a journalist off and on for fifteen years, but in the past, I'd usually worked on the cheap. It's amazing how much difference an expense account makes—now I could follow my subjects for days, a spy with a rental car. After years in the library, I was intoxicated by the possibilities of reporting on the present moment. The wealth of detail dazzled me; I could study their clothes, gestures, knickknacks, voices; I could hang around for days and watch events unfold, accumulating twenty hours of tape. It felt almost like cheating to me, this process of observing, recording, and boiling it all down into a story.

And so I found the doorway to my homeland where I'd least expected it: here in America.

***

No one can agree what to call this chimerical genre, journalism wrapped up in the art of fiction. Truman Capote described *In Cold*

Blood as the first "nonfiction novel." Lee Gutkind, renowned teacher, calls the form "creative nonfiction." My friend Kent Bruyneel—editor of *Grain,* a Canadian literary magazine—has come up with the moniker "not fiction." I've always been partial to the term "true story," because of the simplicity and democracy of that phrase.

Why so many names? I suspect it's because everyone has a different opinion about the truth. For some writers, it's OK to alter minor details in the service of the story; for others, this fudging of the truth amounts to a lie. I spent too many years as a fact-checking drone to be able to cheat the details—I don't fabricate at all. Still, I have to say that great art excuses everything. I understand when a nonfiction author makes a few minor tweaks in service to the scene. When you examine some of Truman Capote's true stories—for instance, the psycho-killer-on-the-loose thriller "Handcarved Coffins"—it becomes clear that he must have smudged some facts, for the story fits together *too* neatly. At the end of the piece, Capote "solves" the mystery and "finds" the killer inside of a fantasy—because, in real life, he was unable to nail the powerful rancher who probably committed the murders. Had he published the piece today, he no doubt would have been called on the carpet by Oprah and forced to account for his lies and inventions. And that would have been a shame, because "Handcarved Coffins" is a masterpiece; indeed, when you read it you feel yourself gripped with a fever dream in which accuracy is simply beside the point.

And yet, Capote was at his greatest when he stuck to the facts. Even the most fertile imagination could not invent some of the dialogue of *In Cold Blood*. Perry, on death row: "I really admired Mr. Clutter, right up until the moment I slit his throat." That's the kind of dialogue you can only bag after spending days listening to someone, and then sifting through your transcripts for the one sentence that contains that character's very DNA.

We talk about certain statements as having a "ring of truth" to them, as if a sentence is a tuning fork, something that we can tap and listen to for its tone. And I think that's right. Truth has a hum to it. You can tell.

***

Several years ago, I found myself in a park in Savannah, leaning on a croquet mallet. I'd flown to Georgia to write a profile of Cheryl Haworth, a 19-year-old who had recently thrown more than three hundred pounds over her head, proving herself the strongest female weightlifter in the world. That day, her best friend Ethan—

a beanpole of a boy—had tagged along. I watched in the shade with Cheryl's mother, who did her best to answer my barrage of questions. At one point, out of nowhere, Cheryl's mom told me that Ethan planned to become a Catholic priest and then work his way up so one day he could be Pope. Right then, I knew I had a story. An aperture had opened up, a chink through which I could peer at the Looking Glass Land of two Southern kids. I had found one of the magic spots where the ordinary melts into the fantastic. Ethan wants to be Pope. Ethan wants to be Pope. My tape recorder hummed in my hand, its reels collecting the evidence. This is what I lived for.

In official terms, I was a "magazine writer." But, really, I'd embarked on a safari, searching out a particular kind of true story. I tracked down visionaries who dared to find solutions to the big problems. These people shared my homesickness in America, and it motivated them to reinvent this country (and others) as a kinder, sexier, smarter, funnier, or more compassionate place. They were possessed of such large ambition that they seemed to violate the very laws of space and time. And, quite often, they managed to bring about the impossible.

For instance, I became fascinated with Gordon Sato, a chemist who had figured out a way to transform the ecosystem of an entire

African country. When I met Sato, he'd already succeeded in creating a three-mile strip of mangroves, enough to furnish an entire village with food. But he was running out of money.

I followed the 76-year-old Sato in the suburbs of Massachusetts as he plotted the future of the African coast, argued with his wife, and ate a few bites of the lunch meat she put on his plate. All the while, I waited to find out why he—a frail old man—needed so badly to transform a desert into a tropical paradise. Eventually I learned that Sato (who is Japanese-American) had been interred in a concentration camp in California as a teenager. Sixty years later, he is still trying to erase the memory of another desert long ago, of a boy—himself—digging in the dust. His outrage is so large as to warp the very fabric of reality. Months after I published my story about Sato, I heard he was sailing to Africa on board a luxury cruise ship with a band of millionaire donors. It was the kind of surreal adventure that seemed ordinary to him.

Some people I profiled in this book are famous. Others live in semiobscurity, each struggling to build his or her paradise out of nothing but sand. Two of the people who appear here went on to win the MacArthur "genius" award. For me, researching these stories felt like scientific observation, an investigation into the nature of personality and reality. How can one man like Sato violate the laws of common sense and live according to his own lights? And how can he convince so many others to join him? Does it come down to American self-invention, this knack we have for making ourselves up out of nothing? Or does the urge come from some more primeval part of the personality? The method of study was simple: I showed up in the right place, and I stayed as long as I could. If possible, I would follow people for days—observing, snooping, asking questions, rifling through their drawers, searching in the cracks between the upholstery in their cars.

And, too, I intended to illuminate and celebrate genius, which always has one foot in reality and another foot in Wonderland. The subjects of these stories consulted the backwards land of their own imaginations, and then made scientific breakthroughs or started new cultural movements.

Most of the stories I wrote under contract to magazines. Two editors—Hugo Lindgren at *The New York Times Magazine,* and John Koch at *The Boston Globe Magazine*—became my champions. They suggested story ideas and provided the enthusiasm that fueled my work. Knowing that I could place these stories in magazines kept me going.

However, the title story, *The Dangerous Joy of Dr. Sex,* was simply too odd to publish in a glossy magazine. I wrote it anyway. When I found out that Alex Comfort, the author of *The Joy of Sex,* had started his career as a virgin and workaholic, I had to know more about him. I wanted to trace his evolution from somber scientist into a pop-culture guru nicknamed Dr. Sex. How could one man change so utterly?

Comfort died in the year 2000, so I couldn't report on him in the usual way; instead, I had to resort to the methods of a biographer. I interviewed his son at length—and here I should thank Nick Comfort for graciously opening up his life to me. I also tracked down many others who'd known Alex Comfort, and read through dozens of books and clippings. In the end, I became fascinated by Comfort's decline rather than his rise—his final ten years trapped in a paralyzed body. That was the kind of plot twist beyond my powers of invention.

As I trailed after my subjects, I was continually amazed by the lines of dialogue that they dropped—their speech so much more poetic than anything I could have made up. And I was awed, too, by the sheer creativity that goes into being human: the anecdotes people think up to explain themselves, their rituals, their plots, their costumes.

In most of the stories collected here, I have tried to open up a doorway between the ordinary and the fantastic. My method was to hang onto the coattails of exceptional people and let them zoom me through magic doorways, into a strange new realm: America as it exists all around us. America, the real.

Pagan Kennedy

March, 2008

Contents

xi . . . . .Introduction

Section 1: Visionaries

1 . . . . .The Dangerous Joy of Dr. Sex

53 . . . . .Genius on Two Dollars a Day

69 . . . . .Bird Brain

77 . . . . .The Strongest Woman in the World

91 . . . . .Battery-Powered Brain

111 . . . . .Vermin Supreme Wants to Be Your Tyrant

123 . . . . .The Chemist in the Desert

135 . . . . .One Room, Three-Thousand Brains

149 . . . . .The Mystic Mechanic

157 . . . . .The Ballad of Conor Oberst

173 . . . . .How to Make (Almost) Anything

189 . . . . .What We Mean By Freedom

Section 2: First Person: Stories from My Own Life

209 . . . . .Boston Marriage

223 . . . . .The Encyclopedia of Scorpions

243 . . . . .Off Season

SECTION 1:

Visionaries

The Dangerous Joy of Dr. Sex:

The story of Alex Comfort, in 17 positions

In 1972, *The Joy of Sex* skyrocketed to the top of the bestseller lists and stayed for most of the decade. It brought the sexual revolution—which had exploded on college campuses a few years before—into the suburbs. Housewives read it and experienced their very first orgasms. Couples pored over it together. Swingers referred to it in conversation with arched eyebrows. *The Joy of Sex* became the Bible of the American bedroom, and it added new terms to our language: *g-string, tongue bath, water works.* Yet, though *Joy* was as much a '70s superstar as Farrah Fawcett, few people can tell you who wrote it. Its author, Alex Comfort, might be considered one of the greatest and strangest minds of the twentieth century.

This is his story.

FIREWORKS

One day in 1934, he sequestered himself in his family's greenhouse in London to perform an experiment. Alex Comfort—then 14 years old—had decided to invent his own fireworks. He ground together sugar, sulfur, and saltpeter, an operation so dangerous that most chemists pour water over the ingredients to prevent a blast. Alex neglected to take that precaution. The container exploded. The roof of the greenhouse blew out. A red-tinted vapor hovered in the air before him. Four fingers on his left hand had vanished, leaving a lump of meat with one thumb hanging off it. He felt no pain. Indeed, he found it thrilling to be blown apart.

Or, at least, that's how he told the story later. Alex Comfort loved explosions, even the one that mutilated him. He never would admit any regret at the loss of his four fingers. As a middle-aged physician, he bragged that his stump could be more useful than a conventional hand, particularly when it came to performing certain medical procedures—exploring a woman's birth canal, for instance.

One thing was clear after the accident: Alex should avoid laboratories, at least until he was older. So he set his sights on literary greatness instead. When he was 16, his father took him on a tramp steamer to Buenos Aires and then Senegal; Alex scribbled notes along the way. In 1938, his final year of high school, he published a little gem of a travel book, titled *The Silver River,* billed as the "diary of a schoolboy."

THE GLOVE

When Alex arrived at Cambridge University, the other students stood in awe of him—a published author! He regarded himself as brilliant but ugly. A reed-thin boy in a tweed jacket, he kept his eyes caged behind glittering round glasses and wore a glove on one hand. "I didn't like to ask him why," said Robert Greacen, who befriended Alex during his university years. One day, when they shared a train car together, Alex removed the glove, and Greacen noticed the stump, but still didn't dare mention it.

The truth was, Greacen had fallen under the spell of Alex Comfort. "Even though we were the same age, he seemed like a man ten or twelve years older than me in ideas, reading and opinion." Greacen decided that Alex was the cleverest person he'd ever met.

Indeed. At age 22, Alex began sparring with George Orwell in the pages of *Tribune;* in rhyming verse, they debated whether Britain should have entered World War II. Alex sneered at the concept of a "good war" and denounced the group-think of the British. He was, already, an anarchist.

SNAIL SHELL

Strangely enough, for one so devoted to free thought, Alex remained a virgin throughout most of his university days. "I was a terribly learned little man. I swotted away at my books," he said later. At Cambridge, he rarely spoke to young women—except on Sundays. Then he broke from his studies to run up the stairs of a Congregational church to join an antiwar gathering, young people in corduroys and tweeds, with bobby-pinned hair and precocious pipes. There he met Ruth Harris and her friend Jane Henderson. They seemed to be opposites: Ruth, a pale girl shrouded in a dark coat, had a submissive air about her; Jane, with an explosion of curls, liked to argue about books. Both girls pined after the boy-genius with a leather glove on his hand, but Ruth confessed her love first. Once she'd spilled out her feelings to him, Alex felt honor-bound to her.

In 1943, they married. They commenced to fumble their way through sex acts, ineptly deflowering one another, then set up house in a tree-lined neighborhood outside London.

Did Ruth realize what she was getting into? Raised by church-going Congregational parents, she aspired to be a social worker, to help the poor and then return home to tea cozies. "My mother was happier in a much more stifling, suburban atmosphere than my father," according to their son Nick.

Everyone knew Alex to be an eccentric visionary, and he behaved like one, making odd demands of his shy wife. One day he asked her to wear her bikini when she gardened; he wanted to be able to peer out the window and watch her bend over their rose beds in nearly nothing. Ruth complied. Soon he wanted more. After the milkman clip-clopped down their lane in a horse-drawn cart, Alex asked Ruth to go out with a shovel and collect the manure left behind, then use it to fertilize their flowers. Could she do this in her bikini? Ruth obediently climbed into her swimsuit and headed out onto the street with the shovel. Many years later, in his famous book, Alex would reveal a taste for bondage; he liked to tie women up. His garden games might have been an early attempt to experiment with fetish-play.

Ruth regarded herself as the long-suffering wife of a great man. She tried not to complain, though the pure force of his intellect wore her out. He followed her into rooms ranting about whatever subject obsessed him at the moment—ballroom dancing, electricity, cell growth, dulcimers, cooking, pacifism, anarchism, utopia. "Holding a conversation with Dr. Comfort is rather like racing after an express train that has already puffed out of the station," a journalist wrote later.

For his part, Alex tried to tamp down the impulses that upset his wife, and it cost him dearly. "I suffered from a severe form of migraine and it produced an intensive depression," he said later, of that period. In order to rein himself in, he resorted to following a well-established British trope: he became an introverted polymath, pottering from one enthusiasm to the next. His son Nick, born in 1946, remembers his father building a television from spare parts and glue-soaked Weetabix. Alex also wired up innumerable burglar alarms, which only went off when they weren't supposed to, emitting inappropriate shrieks.

And he wrote with blazing speed—poetry, novels, science papers, sociology. By 1950, he'd published a dozen books. A medical doctor and biologist, he became a leading authority on snails, that creature that symbolizes the slow and cautious and flabby. It seemed that the younger Alex—the boy who blew things up— had been squashed forever and replaced by a morose intellectual.

"He seemed like a workaholic—only we didn't have that word back them. He was puritanical," said Greacen, who added that the only way to spend time with Alex was to find him in his lab or else tag along to antiwar meetings. The two belonged to a coalition of writers denouncing Cold War hostilities. After their meetings, the writers adjourned to a pub, to huff on pipes and jaw about books. But Comfort refused to join them. Instead of socializing, "he would jump in his car and go home," according to Greacen. "He thought I was lazy. Once he said to me, 'Look Bob, you shouldn't hang around in pubs with people. You'll get no work done.' When I spent two or three hours with him, I'd go away absolutely tired—my head would be filled with all he said about literature and politics. I used to wonder when he slept." Alex Comfort obsessed about poems, political rants, novels, scientific studies—always, he hurled himself into a realm of thought.

Then, in the late 1950s, Comfort developed a new obsession, one as dangerous, in its way, as the gunpowder had been. He couldn't stop wondering about Jane, Ruth's best friend from university days, now a frequent dinner guest at their house. His wife was the kind of woman who shrunk into middle age. Jane, on the other hand, blossomed at age 40. At ease in her big-boned and athletic body, she made only a token effort to keep her lipstick on straight. To hell with propriety! Her curls blew out everywhere, like springs popping from the active gears of her mind. She was as full of ideas as Alex. Instead of marrying, she'd devoted herself to books, and now she worked as a librarian at the London School of Economics. She spent all day around professors; she understood men like him.

He had to have her.

JOHN THOMAS

By 1960, Alex had performed intercourse innumerable times with Ruth, yet he knew next to nothing about sex. So he and Jane studied it together. In the beginning, they snuck around behind Ruth's back, rendezvousing at Jane's flat, which became their laboratory.

Jane, though sexually inexperienced, was just as eager to learn as Alex. She would try anything. She twisted herself into positions and he named them. The Viennese Oyster, the Goldfish, the X. She wanted him to take notes on each contortion, to draw diagrams. They somehow managed to snap Polaroid photos of themselves (one wonders how they hit the button on the camera) to document their favorite positions for later use. They were two intellectuals screwing in every contortion possible, using the full arsenal of their erudition to explode each other to smithereens of pleasure. In bed, they went to Cambridge University all over again, figuring out everything they could about erotic bliss, which was at that time an arcane art. Back then, four-letter words still shocked people. At the drugstore, pharmacists kept condoms under lock and key. *Lady Chatterley's Lover,* banned for thirty years, had finally appeared in print, only to be slapped with an obscenity charge; the prosecutor in the case had asked the men in the courtroom whether this was the type of book "you would wish your wife or servants to read." Sex was not just a private matter—it was clandestine.

Before Jane, Alex had suffered alone, trying to subsist on the weak tea of matrimonial intercourse. But now sex—proper sex— brought him back to life. Indeed, it made him feel so good that he decided he needed a new name. In the privacy of his mistress's flat, he dubbed himself "John Thomas."

That was what Lady Chatterley called her gamekeeper's penis. Of course, the name only applied as long as Alex was with Jane. When he washed her smell off him and went out on the street, he became his usual self again: Alex Comfort, MD, DcS, a married man and snail expert who occasionally appeared on the BBC. Also, a middle-aged man. He sagged a bit these days in the stomach.

His glasses rattled on his face. The situation would be difficult to explain to his wife Ruth. He was an adulterer, like the gamekeeper in Lawrence's novel. It was a terrible thing, to screw your wife's oldest friend, but he didn't see how he could stop.

And here was the funny thing: despite his experiments with Jane, he still knew little about sex. How on earth did one really make a study of it? You couldn't learn anything in the library, because all the guidebooks were hopeless. Just how hopeless? A sex guide titled *The Marriage Art* warned readers that "ineptly arranged intercourse leaves [your clothes] in a shambles, your plans for the evening shot." And what about the dangers of well-planned intercourse? The dull, ritualistic acts between husband and wife? Alex Comfort was one of the first intellectuals to worry that bad sex was a plague across Europe and America.

Right from the beginning, he and Jane aspired to do more than just make love sideways, upside down and backwards, with feathers, toes, and swings. They would change sex itself. The two lovers began working on their own guidebook, cataloging all the positions they had tried. Jane, the librarian, had a genius for organizing information. And so they decided they would go topic by topic, alphabetically. A for Anal, B for Big Toe and so on. Using this principle of tireless cataloging, they created a little homemade book, suitable for passing around among friends. The title: "Doing Sex Properly."

Sex, however, turned out to be too explosive and anarchistic a force to do properly. It spilled out of their guidebook and changed Alex entirely. He wanted to blow up Buckingham Palace, and have the pieces rain down in sparks, fireworks of joy that would set everyone to rutting. He wanted to start the sexual revolution all by himself. So, he went to work writing books that he believed would wake people out of their matrimonial comas, and by the early 1960s, he had set himself up as one of Britain's most outspoken advocates for free love. He'd also come clean to Ruth and worked out an agreement with her: he could continue his affair with Jane, so long as he kept it hush-hush.

Only a few close friends knew that he went back and forth between two wives; to strangers, his call for open marriages seemed to be just another of his outrageous political stances. A professor of a certain age, he appeared to only be dreaming of a sexual utopia— not actually struggling to build one in his own home. And Alex did what he could to foster this reputation as an armchair libertine and nothing more. His 1961 novel *Come Out to Play* tells the story of a doctor and his girlfriend who open a sex school;

though the novel sprang out of his adventures with Jane, he dedicated the book to his wife: "For Ruth, who put ideas in my head." He appeared to be a man besotted with his wife—and only her. Jane didn't like being kept a secret, of course, but she would do anything to hold onto Alex.

NICK

One day in 1958, Alex received a letter from his son's boarding school: parents should give their sons a talk about "personal hygiene"—that is, the facts of life—as the school would not provide such delicate information. Soon afterwards, Alex called his son in for a talk, fidgeting as he explained the business between man and woman. He exuded so much embarrassment that 12-year-old Nick did not dare ask questions. "He did very little to fill in the gaps," Nick remembered later. "I probably heard less talk about sex than the average child." Nick ended up turning to his father's medical books for information, poring over cut-away diagrams of the male and female reproductive organs.

Because he boarded at school, Nick didn't see much of his father during his teenaged years. He had no idea when his father began an affair with Jane Henderson, whom the boy regarded as an honorary aunt. But clearly, something strange had happened. His father seemed to be on TV and the radio all the time now.

In 1962, Alex Comfort participated in an antinuclear protest in Trafalgar Square; the police threw him in jail, along with a cabal of distinguished cellmates like Bertrand Russell. In the morning, most of the other professors sprung themselves by paying a nominal fee. Alex refused to bail himself out and stayed in prison for a month, where he had a lovely time discussing roses with his guards, who were enthusiastic gardeners.

News of Alex Comfort's arrest spread through Nick's boarding school. The other boys taunted the 16-year-old. "Jailbird, jailbird," they called after him. For his part, Nick fumed that his father had chosen to go to prison when he didn't have to—just to get attention. Raised by his mother to value moderation, Nick abhorred his father's outrageous style.

In 1963, his father published a book that advocated for free love, *Sex in Society,* and went on the radio to push its radical agenda. He called chastity a health problem. He argued for legal prostitution. He celebrated cultures in which children had sex with one another. He lobbed his ideas like bombs, and the British public, duly scandalized, responded with scathing attacks on him.

"Being at boarding school when that book came out wasn't the best place to be," according to Nick. Back then, as a teenaged boy, he endured taunts for the actions of his faraway father. Seventeen-year-old Nick seethed over one particular remark his father had made in the book—that 15-year-old boys should be given condoms. Trapped at a blazers-and-neckties school, Nick hardly ever saw girls; he blushed and grew sweaty when he tried to talk to females at a school mixer. The last thing he needed was a condom. His father was busy running around decreeing what was best for theoretical teenagers, but he didn't even know his own son.

The truth was, Alex didn't have much left over for Nick. When you reconstruct his schedule in the mid-1960s, he appears to be comically super-human, like one of those lechers in the old Benny Hill TV show who scooted around in fast-forward, their legs flailing as they chased after naked women. Alex would dash home for dinner with Ruth and sleep there; then in the wee hours of the morning, he'd hurry to Jane's flat, nuzzle with her for a while, drive her to her job at the library, then head off to his own duties at University College London. "For quite a large part of my time, I was with two different women," Alex would comment later. "It is not an arrangement I recommend."

Nick tolerated the situation, too, because for the first time ever, his father seemed happy. "There were sparks of levity starting to come through. [My father] felt he'd been too serious when he was young." Now, at last, Alex had discovered his own impishness. His joy.

Being married to two women could be exhausting—the driving alone wore him out. By the late 1960s, he had divvied up his week: Mondays, Wednesdays, and Fridays belonged to Jane; the rest of the week he spent with Ruth. Neither of the women had other lovers, so he had to keep both of them satisfied. His wife would tolerate anything, so long as he played the part of her devoted husband. Nick believes that his mother agreed to the arrangement because "it gave her more time. My father could be such a handful." But one senses that Ruth had simply learned to make the best of a bad situation. If her husband decided to be a free lover, well, she would just bear up and smile. Jane, meanwhile, wanted more of him. He commuted between them in a tricked-out van with a condom taped on the dashboard.

Even many of Alex's friends knew nothing about the Monday-Wednesday-Friday side of his life. Marilyn Yalom—a feminist scholar who had befriended both Alex and Jane—was one of the friends who did know. "He was proud of the fact that he had two women and that he'd arranged his life this way," according to Yalom. "There was something ineffably childish about Alex."

JANE'S UNDERPANTS

In the late 1960s, Yalom was passing through London and fell sick with a high fever. While she recovered, she stayed in an extra room in the flat that Jane and Alex shared. She found the bond between Alex and his "second wife" rather touching. But Yalom had no desire to join in. "When I got better, they proposed a threesome to me. They were like two little children saying, 'This would be fun.' I brushed it off without taking them too seriously," Yalom recounted later. "When I left, packing my things, Jane brought me a pair of her underpants. It was for her some kind of symbol of what I'd missed, or of some kind of intimacy between them."

By now, even though Jane and Alex had both grown wrinkled and so podgy they wore stretch pants, "they were like a couple of teenaged kids," she said. All the while, they had kept on writing their book together, adding exotic positions that they learned about on their travels (sometimes with Ruth in tow) to India and across Europe. They also began to write longingly about group sex:

"Subject of a whole cult today, of which we aren't members, so we speak from hearsay. It's becoming socially more easy to arrange, and personal taste apart we see no earthly reason why pairs of friends shouldn't make love together."

By the late 1960s, Alex realized that his hobby with Jane might turn into an important book indeed; he believed they had compiled and documented more sexual practices than anyone else in the world, with the possible exception of the Fourth Century scholar who edited the *Kama Sutra.* So one day, Alex scheduled a meeting at Mitchell Beazley, the UK's top publisher of illustrated encyclopedias, to show off the manuscript. The editors recognized the book as a masterpiece, the most comprehensive book on sex ever written, and every position in it "kitchen tested." They'd take it.

THE DANGERS OF STICKING WITH SNAILS (an aside)

"It is an iron rule in our trade that no one gets anywhere until they give up snails," biologist Steve Jones has observed, pointing out that Edgar Allen Poe, Lewis Carroll and Alex Comfort all achieved greatness only after they lost interest in mollusks.

By 1969, Comfort had spent two decades pottering around in labs, surrounded by piles of snail shells and plastic jugs full of guppies; he'd been searching for the mechanism that causes mollusks and fish to grow old. He'd first become interested in the tick-tock of death back when he was a young man. In his 20s, as a poet-scientist sickened by the needless destruction of Hiroshima and Dresden, he'd belonged to a literary movement called the New Apocalyptics. In his poetry and in his lab, too, he'd explored the clockwork of decay. But now, he'd become a different sort of person entirely. A swinger. A sex expert. A happy man. Death no longer interested him.

Picture him in his office in London, packing shells in cotton and labeling boxes. He takes apart his chromoscope and drops the pieces into a box. Soon he and Jane will fly to Los Angeles, to the blinding sun and the dry canyons where few snails crawl. There, he will study another kind of specimen. He has heard about a utopian community out in those hills, where couples are experimenting with orgies. He will finally get a chance to participate in a laboratory of free love. Alex Comfort is about to drop snails forever and go on to bigger things.

DEDICATED TO NO ONE

Meanwhile, he finished up work on his sex manual. As he did so, he began to worry about the risks of publishing what was, in effect, a diary of his sex life with Jane, written in the voice of a couple: "we" prefer playing in the bedroom, "she" likes her nipples licked, and so on. Had Alex owned up to being the male half of the couple, most readers might assume his wife Ruth to be his toe-riding, rope-loving, corset-wearing partner. Or the audience would suspect the truth: that his coauthor was his mistress. Either way, the book would destroy his marriage. It might also lead to the suspension of his license to practice as a doctor; the British medical board had been known to punish MDs who wrote scandalous tracts.

In the end, he tried to get around such problems by listing himself as the editor of the book, rather than its author. He lied in his introduction, saying that he had gathered his material from other people rather than from first-hand experience. "The book is based originally on the work of one couple. One of them is a practicing physician; their anonymity is accordingly professional," he explained. He also pretended not to have written the text, saying he had "left alone" the "authors'…light-hearted style." Later on, he told journalists that he'd resorted to this subterfuge in order to protect his reputation as a doctor. No doubt Alex also hoped to spare Ruth the cruelty of going public with his sex life. After all, his marriage to Ruth had endured through the wild 1960s, and though he saw less of her in 1970, he still loved her.

He had dedicated almost every one of his previous books to his wife. This time, he included no dedication page at all.

THE BEARDED MAN

The original title for the book, *Doing Sex Properly,* sounded too stern, so an editor suggested a better one: *The Joy of Sex* would capitalize on the popularity of Erma Rombauer's bestselling cookbook

The idea of treating sexual play as if it were French cooking delighted Alex. His mother had been the first woman ever to win a scholarship to Oxford University in French studies; when Alex was born, she dropped out of school and concentrated on sculpting him to be a perfect little wunderkind. By the age of seven, Alex could speak French fluently. The language still came easily to him, and he scattered it all through *The Joy of Sex,* like so much gold dust. Foreign phrases worked magic. Ass licking became *feuille de rose.* Sniffing the partner's armpit became *cassolette;* a finger up the anus, *postillionage.* He made acts that were illegal in many countries sound like rare, sought-after perfumes. Without quite realizing what he was up to, Alex had found a way to re-brand sex. He'd taken stunts seen nowhere outside of triple-X porn and turned them into entertainments suitable for a suburban couple to try after a few glasses of merlot. Looking back, his greatest stroke of genius was to abandon the one-size-fits-all model used in nearly all the previous sex manuals. Alex—himself a bondage and group-sex fan—championed the idea of different tastes for different tongues. Today, with the Internet at our fingertips, we're used to people dividing themselves up into erotic tribes: costume wearers, horse-play enthusiasts, and dungeon mistresses have found each other and formed communities. But in 1970, few people had heard of such variations. Alex treated it all as just part of the *pot-au-feu* of human desire, and his tone suggested that the reader would too. He imagined that *Joy* would become a "coffee table" book; on display in the living room, it would make an excellent ice breaker, and might even encourage the Joneses to invite their neighbors the Smiths for a foursome.

At the same time, he recognized that many of his readers would be sexual imbeciles. The book included a sewing pattern for a g-string, so that readers could make their own stripper clothing and wear it around the bedroom. What's mind-boggling now, decades on, is that g-strings were not available in any store back then. Imagine the housewife who cut out the pattern, shopped for black silk, and painstakingly hand-sewed the adornment, all so she could surprise her husband. In 1970, when Alex prepared his sex tips, most adults were married. The birth control pill had been available for less than ten years. Many of Alex's readers had wed at 20 or 21, and had slept with only one partner. They were poignantly aware that they could be better lovers.

The book reflects these aspirations. Follow instructions, its tone implied, and you could develop the kind of exceptional lovemaking skills previously available only to the rich. Here were the tricks you might learn if you spent a few years bumming around Europe, jetsetting to nude beaches, dabbling with lovers who spoke French. World-class sex required more than just practice—you had to learn from the experts. "Chef-grade cooking doesn't happen naturally," Alex exhorted his readers. "It starts at the point where people know how to prepare and enjoy food, are curious about it and willing to take trouble preparing it, read recipe hints, and find they are helped by one or two detailed techniques. It's hard to make mayonnaise by trial and error, for instance. Cordon Bleu sex, as we define it, is exactly the same situation."

It was a message that would resonate with millions of people— their sex could be thrilling and also classy.

But what about the illustrations? *The Joy of Sex* would describe positions that few of Comfort's readers had ever encountered before—Chinese style, Indian style, croupade, cuissade. Obviously, it would be necessary to include pictures that showed how to insert part A into slot B. Alex handed over the Polaroid photos that he and Jane had snapped years ago to document their adventures. Couldn't artists turn these photos into sexy drawings? The smudgy, dark Polaroids of two middle-aged intellectuals smashed together in various tableaus were far from appetizing. Both Charles Raymond and Chris Foss, the men assigned to illustrate the book, decided they must go looking for inspiration elsewhere. They leafed through porn magazines and hired models, but they couldn't seem to get the right look. Furthermore, the editors at Mitchell Beazley nixed the idea of photographs, which would be likely to get the book banned.

In the end, everyone agreed *The Joy of Sex* should be filled with line drawings; the softness of the pencil would add a homespun intimacy and let the book slip past the censors. The drawings would be inspired by photographs of a real-looking couple—though the models would not be as real and saggy as Alex and Jane.

One of the illustrators, Charles Raymond, volunteered himself and his wife Edeltraud. Charlie had a hippie beard and the long, tangled hair of a wino, but he turned out to be well-suited for his new role as sex model. Chris Foss, who snapped the photos, could only marvel at the ease with which Charlie and his wife bent themselves into the 200 positions they documented for the book. "Edeltraud was very Germanic and so she would say 'Right, Charles, we start now. Position number one!'" remembers Foss. "She'd tap her leg and say 'Come on, Charles!' and off they would go and do that and she would tick it off and say 'Right Charles, now we do this one!' Poor old Charlie was only human, so every now and then he—how do I say it—blew it a bit. I'd say 'Charlie, you can't do it now. We've got 15 more positions and just an hour left. Charlie, can you get back into business as quick as you can?'"

Afterwards, Foss produced minimalist line drawings based on the photos, gorgeous in their simplicity. His illustrations capture the many moods of two people who adore each other. In some, the lovers droop across one another, satisfied; in others, they rear back, trying to find just the right spot. But, lovely and erotic as the illustrations are, the first thing you notice about them is Charles Raymond's hair, that wild mop on his head and the obscene beard.

When the book came out, much of the buzz would center around the "bearded man," who seems to be the narrator and hero of the story. His caveman coif and perpetually hard penis would do as much to sell the book as Alex's elegant prose.

Of course, as Alex worked on his manuscript in 1970, he had no idea just how popular the book would become two years later. The publisher planned an initial print run of only 10,000 copies. For Alex, *Joy* was a manifesto, a byproduct of his own revolutionary sexual life. He had begun commuting back and forth to the United States to spend weeks at a time living in a community called Sandstone. He had just discovered an erotic playground that could only have existed in his wildest fantasies—or in California.

SANDSTONE

"The moment [my father] went to the States, the party began," according to Nick Comfort. "He had a suffocating upbringing. When he discovered what life could be like in California there was no stopping him." *Joy* had yet to be released, but already Alex had embraced his new role as the run-amok British intellectual. In 1970, he and Jane flew to Los Angeles, to get their first taste of Sandstone.

Most of the couples that made the trip to the Topanga Canyon community went in search of utopia. Like Alex and Jane, they believed Sandstone offered the first glimpse of what society would be like in the future, when Americans built their lives around pleasure rather than duty. Founders Barbara and John Williamson handpicked the members, seeking out professors, journalists and scientists who might show up at one of the naked dinners, stay for the orgy, become a convert to the philosophy, and then spread the word in newspapers and academic studies. Their schmoozing paid off. George Plimpton, Dean Martin, Tim Leary and Daniel Ellsberg were just some of the pop luminaries and intellectuals who rubbed up against each other, perhaps literally, during weekends. Sammy Davis Jr. materialized one night in full Vegas drag—diamonds on his cuffs and cigarette lighter—escorting both his wife and porn star Marilyn Chambers. As he stripped, the cufflinks bounced to the floor and Chambers dove for them.

But this was no mere sex club. Only couples could join the Sandstone community, and everyone had to abide by a set of rules. The Williamsons saw Sandstone as a revolution in lifestyle: they wanted no less than to redesign commitment between men and women, to teach husbands and wives to put aside "hang ups" and give each other complete freedom. It was a have-your-wedding-cake-and-eat-it-too philosophy of love: men and women were supposed to be able to screw anyone they fancied, and do it on a mattress in front of their spouses; this would (in theory) deepen their trust in one another. Alex himself couldn't have designed a better test-tube for his ideas.

One evening in 1970, he maneuvered a car up the Pacific Coast Highway, heading to his first orgy. The road to Sandstone was marked only by discreet stone pillars, and the visitor had to bump along a private drive until he reached the main house.

That evening, Alex parked next to a fleet of Jaguars and Porsches, and climbed out into the smell of eucalyptus, pot smoke, and suntan lotion. At the door, Barbara Williamson, a slim woman in her 30s wearing nothing but spectacles, checked his name against a list. The house was a 1970s California suburban spread—redwood deck, low-slung furniture, wagon-wheel chandelier. The wallpaper, in various shades of brown and mushroom, appeared to be paisley until you got up close to it; then you saw the naked bodies worked into the design. A sign hung above the garlands of incense, informing all visitors about the rules inside these redwood walls:

Privacy means you see the environment as hostile….Here, there are no doors on the bathrooms, the environment is gentling. Lovemaking is…tribalized. You are in the community. You do your thing; it is yours. Others may want to watch, learn, comment, even laugh. But they are of your tribe; the laughter is not hostile.

Once inside, Alex was free to lounge by the fire in the living room and chat with other naked professionals. Or he could descend the stairs into the "ballroom," a wall-to-wall carpet of seething flesh: breasts, butts, astonished faces, hair, writhing legs, errant arms, coos, sighs, slurps, giggles. Later, he would write about those Sandstone parties as mind-expanding experiences akin to LSD trips. Usually, the woman ended up being the more enthusiastic half of the couple, he remarked, and led the way downstairs into the orgy room; because of her superior anatomy, she could keep going all night, long after her husband was sidelined. In the morning, the first-timers tended to go on talking jags, as they tried to hold on to revelations from the night before. They would linger over their coffee, "trying to work out what had happened to them," Alex wrote. "Breakfast often ended as an impromptu seminar on something," as the Sandstone-ites rapped about how to take their insights to the wider world. "Some of these morning-after experiences were even more rewarding than the specifically sexual part of Sandstone," he wrote.

Alex fell in love with the community. From 1970 through 1972, he lived there for weeks at a time. Occasionally Jane accompanied him—for they were still very much in love—but often he pilgrimaged to Sandstone solo or with a male buddy.

It seemed that Alex could break the rules. Gay Talese described Alex as a fixture at the place, its eminence grise. "The nude biologist Dr. Alex Comfort, brandishing a cigar, traipsed through the room between the prone bodies with the professional air of a lepidopterist strolling through the fields waving a butterfly net," Talese wrote in *Thy Neighbor's Wife.* "With the least amount of encouragement—after he had deposited his cigar in a safe place—he would join a friendly clutch of bodies and contribute to the merriment."

It was at Sandstone, and often in mid-orgy, that Alex met many of the other minds behind the sexual revolution. For instance, Betty Dodson—the feminist writer who championed masturbation in her own bestselling book—first noticed Alex across a crowded room at Sandstone. At the time, Dodson happened to be stretched out on the floor with a man and a woman licking her. Looking up from the proceedings, she saw a gent with gray curls who studied her intently. She waved him over, and he tottered toward her obligingly. When he was near enough, she reached for his left hand, as if to shake it, and found nothing there but a stump and a thumb. Unperturbed, she slapped his hand down on her crotch. "Oh, he was so happy," Dodson remembered years later.

But it all came at a cost. Alex felt he had to cut himself off from his family and old friends, and even his son. "He was always strongly resistant to the idea that I or any of the family should visit him in California—largely because he didn't want us privy to the Sandstone community and all the rest," according to Nick. Father and son had no contact for years. The relationship "was very much a one-way thing," according to Nick. For all Alex's talk of freedom from shame, he was too mortified to let Nick witness the goings-on in California.

DR. GOGGINS

When *The Joy of Sex* came out in 1972 it hit the bestseller lists in most English-speaking countries. However, its success was most blazingly spectacular in the United States, where it climbed to the top of the charts, stayed at number one for more than a year, and remained a bestseller for six. "Nothing makes a recreation more respectable for Americans than the rumor that it takes know-how," the critic Hugh Kenner would later quip about the book. Alex's writings fit perfectly with the American zeitgeist: he'd presented sex as wholesome sport, sanctified by puritanical labor and an effortful sheen of sweat.

Even as the book became a superstar, Alex himself remained in the shadows. He might have exploded into fame as Timothy Leary had, might have become a walking advertisement for his own brand of utopia. Except, of course, Leary possessed the movie-star jaw, the Pepsodent teeth, and the boyish physicality that allowed him to transition from professor into pop-culture icon, and Alex...didn't. Had you met him in 1973, you might well have mistaken him for a car mechanic. He made the rounds of newspaper interviews in a blue jumpsuit—no need to wash shirts, he liked to brag, if you only wore zip-up suits. He slumped in his chair with a hunched-over posture that made him look older than his 53 years, and kept his breast pocket stuffed with a line of cigars, and another stogie clamped in his mouth. Photographs of him during this period suggest that he washed his gray hair as infrequently as his costume.

And anyway, Alex did not crave fame of the Leary variety. Rather, he saw himself as the man who lobbed the bomb, caused the big explosion that changed society, and then ran away. An avowed anarchist, he lived for the moment when the old system fell apart and the new order was born.

Ever since the early '60s, he'd imagined himself in the role of world-class trickster. In his 1961 novel *Come Out To Play*, he had created a character—Dr. Goggins—who represented his ideal. A genius in bed, the doctor opens a school in Paris where he and his girlfriend teach their students advanced love-making skills. "Bill Masters thought I was writing about his clinic, but I was able to tell him I wrote that book back before Masters and Johnson were heard of," Alex bragged in 1974. And now it appeared that what Alex had imagined a decade earlier might come true.

In the novel, Dr. Goggins meets "a French chemist who's happened on a drug [that] can turn people on. Not raise the libido, but thaw the superego, the part of the mind that says 'mustn't.' I call [the drug] 3-blindmycin," Alex explained in one interview. Dr. Goggins—believing that enough orgasms could bring an end to war—constructs a Molotov cocktail, fills it with a huge dose of the drug, and sets it off in front of Buckingham Palace. The queen, parliamentarians, barristers, peers and bureaucrats all huff in the substance and become flower children. The military brass retire to the country. The rocket scientists lose interest in building phallic weaponry and run naked through their neighborhoods. War ends forever.

"Come Out To Play started to be simply a comic novel. I think now it was the manifesto of which *The Joy of Sex* commences the implementation," Alex told a journalist. After 1972, he tried hard to sell *Come Out To Play* to Hollywood—with no luck. Still, he could imagine exactly how it would look on the big screen. Peter Sellers would star as Dr. Goggins, the ultimate prankster, a man who brings about true democracy by wiping out the will to power. In the early 1970s, Alex could believe that he had himself become the Peter-Sellers-as-Goggins character; *Joy* was his drug, his 3-blindmycin. With the right words, he had transformed the mores and habits of the entire English-speaking world. He'd turned what used to be called perversions into "pickles and sauces"; he'd written the manifesto that people were using to refashion their lives, putting the orgasm at the top of their agendas.

Up until now, Alex had been out-of-step with the mainstream; in the 1940s, his countrymen denounced him as soft on the Nazis; in the 1960s, moralists had crusaded against the radical agenda he put forward in *Sex and Society*. Now, for the first time, he led the crowd.

THAT BOOK

The runaway success of *Joy* presented one problem for Alex: he'd made Ruth's life hell. Every one of her acquaintances, from grocer to next-door neighbor, had likely heard about her husband's exploits with Jane. And though Alex had gone through gyrations to pretend he was only the editor, that subterfuge evaporated after the book came out. The press called him "Dr. Sex." Ruth coped with this disaster by pretending it wasn't happening. According to Nick, "my mother never discussed [*Joy*] at all." When she absolutely had to allude to her husband's bestseller, she called it "that book." The marriage, which had survived so many betrayals, now crumbled.

In early 1973, Alex negotiated a divorce from Ruth. A few months later, he and Jane married in a quiet ceremony in London. Nick—and many old friends—were not invited. Then the couple jetted off to their new home in California. The Center for the Study of Democratic Institutions, a liberal think tank in Santa Barbara, had offered Alex a sun-drenched office near the beach. He'd take it.

GOOD HOUSEKEEPING

"Here we have a new genre: the coffee-table book that should be kept out of the reach of the children. Higher coffee tables would seem to be the answer," John Updike joked in the 1970s, about *The Joy of Sex*. But the coffee tables stayed low, and the book splayed its pages suggestively in living rooms across the nation. It matched the Danish Modern furniture; the understated cover was all white space, the epitome of space-age minimalism. The book became a fashion accessory, a symbol of the New Good Life. Along with a waterbed, a hot tub, and a high-end stereo, this was a toy for adults at play.

And suddenly, so many were at play: in 1970, California adopted the first no-fault divorce law, and by the end of the decade splitting up had become easy. Now, entire consumer empires catered to the single and middle-aged: these swingers needed fern bars, sports cars outfitted with quadrophonic 8-track tape machines, amusing drinks (Harvey Wallbanger, Harvey's Bristol Cream), aftershaves, designer jeans, beachside condos, shag rugs, silver coke spoons.

Alex Comfort had never meant to create a faddish product or storm the consumer market. Just the opposite: he dreamed of a utopia where tribes of people shared their bodies, where greed dissolved. In such a world, talking about money would be the ultimate gauchery. Examine the illustrations in *Joy*, and you catch a glimpse of what heaven might have looked like in Comfort's mind. The lovers float in the freedom of white space; seemingly they own few possessions beyond pillows and a mattress. They need nothing but their erogenous zones. In the sequel, *More Joy*, their friends come over, and now the foursome copulates on some hazy surface. (The floor? A bed?) The reader of *Joy* steps into a demimonde something like Sandstone in the early years—just men, women, and mattresses. Pleasure belonged to everyone. Didn't it?

The last thing Alex Comfort had expected was for his guidebook to become a status symbol. The book made a fortune for its publisher and became a must-have furnishing for the vacation house or bachelor pad. It both promoted Alex Comfort's utopia and undercut its essential message: in *Joy*, he exhorted readers to get beyond all the pre-packaged ideas about sex, and to fully inhabit their own minds and bodies. "Play it your own way," he lectured readers. Once "you have tried all your own creative sexual fantasies, you won't need books." Ah, but his readers did need the book; they bought it by the millions; they wanted the cool white cover; they longed for an artifact.

Likewise, his beloved Sandstone became an upscale consumer product in the late 1970s. Its original owners, bankrupted by a lawsuit, had no choice but to sell the place. The new proprietor, an ex-Marine, instituted a $740 initiation fee for members, which made the club as expensive as a Hawaiian vacation or a share in a yacht. Once Sandstone turned into a money mill, Alex divorced himself from the place. By all accounts, he gave up on group sex entirely around that time.

MONOGAMY

But wait! Let's not go there yet. Let's jump back to 1972, to the happy times. In those first golden years in California, Alex and Jane fucked and recited limericks and threw dinner parties. They had set up house outside Santa Barbara, in Montecito, and seemed destined to slither into old age bathed in massage oil. They padded around naked and floated in swimming pools as blue as Viagra—though, of course, Viagra hadn't been invented yet. The sex pill of choice was still 3-blindmycin, Alex's imaginary aphrodisiac.

At first, Jane "was willing, maybe even eager, to participate in the sexual play," according to Marilyn Yalom. Alex boasted about her prowess in bed, and even created a stage on which both of them could perform; in the back of their house, he installed an "Indian room" filled with batiks and pillows where he could throw orgies. "He assumed that the two of them would be the sexual gurus of Montecito," according to Yalom. "But Jane balked."

Stranded in an alien country, Jane began to regret indulging Alex's every sexual whim. In the late 1970s, she told her husband she wanted a conventional marriage—no group scenes, no experiments, no second or third wives. "I can still see the look of disappointment on [Alex's] face," Yalom remembers. "He had more or less contracted for a sexual playmate where the doors would be open. She changed her mind."

Alex had believed that open marriage and group sex would become the model around which most people organized their lives—and he'd been wrong, on both a personal and a cultural level. According to Yalom, "He didn't realize the strength of monogamy."

In 1985, he and Jane decided to move back to Britain. Jane would feel at home there, and Alex might be able to resume serious academic work. Besides, they were now well into their 60s. "[My father] came back to the UK to enjoy his old age," according to Nick. "But that didn't happen."

AGE-ISM

In the late 1970s, around the time that Jane demanded his fidelity, Alex's magpie mind led him to return to a question that had fascinated him as a young man: how and why do people grow old?

Since the 1940s, he'd been reading and publishing papers on gerontology; now he hoped to do for the aged what he had once done for the sexually innocent. In 1976, he published a manifesto against the warehousing and dehumanizing of old people, titled *A Good Age.* He intended his new book to be as groundbreaking as *Joy,* a bestseller that would radicalize the wrinkled. However, as it turned out, readers were not as eager to join this revolution. The book sold respectably, but couldn't begin to match the influence of Comfort's big hit. Still, Alex kept on with his crusade. He flew to conferences on gerontology, trying to wrangle his way to the top of the field. When the proceedings bored him, he jumped up on stage and summarized the disagreements with a limerick, written on the spot. He couldn't resist proving how much smarter he was than any of them. The scientists and doctors out there in the audience, those men with their sober expressions and sharp name tags, thought of him as a pop-culture idiot. *Joy* had ruined his reputation among serious people. He wanted it back.

He shed his jumpsuits for tweed blazers; he anticipated the breakthroughs he'd make in the field. He published a shower of books and articles. But wherever he turned, he was dismissed as the guy who had written that sex book.

"There was the publicity roller-coaster which he quite enjoyed," according to Nick Comfort. "But he wanted to be treated seriously. He was an academic of considerable gravity." Still, it wasn't just his pop-culture past that tainted his reputation as a biologist. Alex had little interest in making a scientific argument; instead, he published polemics. His crusade had been—and would be—to convince people to see their bodies as ideas. This may be one reason that he did not anticipate AIDS, even in the midst of the orgies, with fluids splattered everywhere. Alex could not seem to accept that viruses and cells and bodies themselves might have antirevolutionary agendas.

Likewise, he insisted that most of the problems of the old came from discrimination—rather than decay or illness. And now he began to imagine a new kind of utopia, a place where the gray-haired could be cherished as sexual and intellectual beings. In such a society, he believed, health problems would nearly vanish for the old.

THREE STROKES

It was 1991. The house in Kent, England, smelled of curry and pipe smoke. Prints from India—which he and Jane had collected during their many trips—hung above the sofa. Alex had embraced his role as the globe-trotting gadfly; he jetted around, speaking out against age-ism.

Today, he was pottering around in his study, working up notes for a nursing conference in Australia. Except he couldn't seem to concentrate. His thoughts had gone foggy. He picked up a pencil to make a correction on his manuscript. Australia. He'd never been before. He was just entering his 70s, and by his calculations, he had twenty good years ahead of him, at least. The rest of his life stretched ahead, enormous as Australia, undiscovered and sunny.

And then he collapsed.

It was Jane who found him. She punched numbers into the phone. Soon, paramedics crowded around, snapping their latex gloves, prodding him all over. He rode in an ambulance and then a helicopter. The next day, he woke up in a new country; it was not Australia. The monitors beeped beside him. The nurses—whom he would never again lecture in conference halls—whisked in and out. His jaw had frozen up. He shook all over. The doctor came by and explained—slowly, as if Alex were an idiot—that he'd suffered a massive brain hemorrhage. He'd obviously taken a hit to his motor cortex. Years before, writing *A Good Age,* he'd been so confident: "The human brain does not shrink, wilt, perish or deteriorate with age. It normally continues to function well through as many as nine decades," he'd asserted. But now his own brain begged to differ.

He drifted in and out of consciousness for weeks. Someone held his hand and told him Jane had died. She'd suffered a massive heart attack.

Seven months later, they put him in a wheelchair and his son Nick wheeled him out into the world of taxicabs and crowded crosswalks, where everyone had somewhere to rush. Everyone but him. Nick—the son he'd hardly known—would handle most things for him now, from bills to fan letters to updating the next edition of *The Joy of Sex.* He would be installed in a little flat around the corner from where Nick, his wife, and the grandchildren lived. And do what? "Leisure is a con," he'd written, but now he had nothing but idleness ahead of him. He would enter that condition he feared most, and which he'd tried to eliminate with his theories: he'd become an unperson.

When the next stroke hit him, paralyzing his right side, he moved into a nursing home. On Sundays, he rode in an ambulance to Nick's house and sat immobilized while his grandchildren ran around him. When he spoke, people leaned in, wrinkled their foreheads, and asked him to repeat himself.

ALBATROSS

The book had made him a multimillionaire and vaulted him to worldwide fame. But he didn't want to be famous like this, a wax figure propped up in a chair, expected to talk endlessly about the same goddamn thing. "*Joy* is frankly an albatross," he said to one journalist in 1997, pumping his thumb for emphasis.

In his mid-70s, Alex Comfort had shrunk into a codger with Einstein hair and furious eyes, immobile in his wheelchair. With his right hand paralyzed, his left thumb had become his only lever on the world. He typed poetry with it, one painful letter at a time. He gestured by wiggling that thumb, so that his whole will seemed to live in that snail-like appendage.

Today a young reporter—a girl, really—would be visiting him; she'd told him she was doing a story about the 25th anniversary of

Joy. Before the girl came, his nurse, Linda, set him up in a spotless suit and arranged him in the chair.

The reporters always asked the same question. How did he come to write *The Joy of Sex?* And behind their question lay another question that he heard in the snigger of their voices. They cared nothing about the other 50 books he'd written, now arrayed on shelves behind him. Always *Joy, Joy, Joy.* Had they not read his poetry? His scientific papers?

The reporter—a beautiful girl, just as he'd hoped—entered and reached out for his hand. But he had nothing to shake, so she smiled awkwardly. She asked where to put her tape recorder. "Stick it wherever you like," he tried to quip, but his mouth wouldn't cooperate.

The girl questioned him for hours, and because she was kind enough to flirt a little, to make him feel human for a moment, he did his best to answer.

She wanted to know whether he'd marry again.

He laughed. Who would want a cripple stuck in a wheelchair?

But your brain works, the girl said.

Yes, that's true, he replied. But no one wants to marry a brain.

He wanted to tell her it wasn't supposed to go like this.

He thought he'd known his enemies: the Tories, Thatcher, the sword rattlers, the physicists who designed bombs, the prudes, the teetotalers. But now he'd been defeated in the way he least expected. The true enemy had turned out to be so small, and not at all political: bits of brain tissue in the motor cortex and possibly the parietal lobe. And *Joy* had become an enemy too. The book was an albatross straight out of Coleridge.

Remember how the poem goes? A ship gets lost in a storm, and bangs around among ice floes; the sailors believe they'll die at sea. Then salvation comes in the shape of a bird, an albatross that leads them to open water. The sailors venerate the bird as an emissary from God—or most of them do. One free-thinking anarchist on board refuses to acknowledge the bird as holy. This anarchist, who loves shocks and explosions of all kinds, shoots the albatross dead.

And then the winds cease, the sails drop, and the ship founders in the terrible silence of the sea. The other men hang the dead bird around the anarchist's neck as a sign of his sin. Years pass. The anarchist recants. And yet, still he wears the ghost of that dead bird around his neck. It weighs him down. It humbles him. And eventually he and his greatest mistake merge into one.

Alex Comfort died in 2000.

SOURCES

Alex Comfort obituary. *The Guardian* (London), March 28, 2000.

Allyn, David. *Make Love, Not War: An Unfettered History.* New York: Routledge, 2000.

Banks-Smith, Nancy. "Privates on Parade." *The Guardian* (London), Jan. 5, 2001.

Comfort, Alex. *Come Out To Play.* New York: Crown Publishers, 1975.

———. *A Good Age.* New York: Crown Publishers, 1976.

———, ed. *The Joy of Sex: A Cordon Bleu Guide to Lovemaking.* New York: Crown Publishers, 1972.

———, ed. *More Joy.* Crown Publishers, 1974.

———. *The Silver River: Being the Diary of a Schoolboy in the South Atlantic.* London: Chapman & Hall, 1936.

———. *Writings Against Power & Death.* London: Freedom Press, 1994.

De Bertodano, Helena. "25 Years of Joy." *Chicago Sun-Times,* Jan. 5, 1997.

———. "Labour of Love." *Courier Mail* (Queensland, Australia), Nov. 23, 1996.

Dery, Mark. "Paradise Lust." *Vogue Hommes,* Spring/Summer 2005, pp. 244–7.

Gray, Paul. "Less Joy." *Time,* May 19, 1975.

Hammel, Lisa. "Will the Real Alex Comfort Please Stand Up?" *The New York Times,* June 2, 1974.

Heller, Zoe. "Where's the Beard?" *The Daily Telegraph* (London), Sept. 17, 2002.

Hunt, Liz. "Alex Comfort's Joy of…Poetry." *The Independent* (London), Nov. 12, 1996.

Illman, John. "Has the Joy of Sex Stood the Test of Time?" *The Guardian* (London), Nov. 12, 1996.

Jones, Patti. "Better than Sex." *The Seattle Times,* Jan. 27, 2003.

Jones, Steve. "Adventures in Thought: Experiment in Literature and Science." Transcript of a talk given at the University of East Anglia on June 16, 2006.

———. "My Greatest Mistake." *The Independent* (London), Sep. 3, 2003.

Kenner, Hugh. "The Comfort Behind the Joy of Sex." *The New York Times Magazine,* Oct. 27, 1974.

Martin, Douglas. "Alex Comfort, 80, Dies…" *The New York Times,* March 29, 2000.

Salmon, Arthur E. *Alex Comfort.* Boston: Twayne Publishers, 1978.

Savage, Dan. "The Doctor of Love." *The New York Times,* Jan. 7, 2001.

Sayre, Nora. "Ahh!" *The New York Times,* Feb. 11, 1973.

Stuever, Hank. "Joy of Sex Back on Top?" Author's Son Updates a 30-Year-Old Classic." *The Washington Post,* Feb. 25, 2003.

Talese, Gay. *Thy Neighbor's Wife.* Garden City, New York: Doubleday & Co., 1980.

Twenty Twenty. "The Joy of Sex." First aired on BBC Channel 4 in 2001.

Updike, John. "Coffee Table Books for Higher Coffee Tables." *The New York Times,* Oct. 28, 1973.

Walsh, John. "Everything We Wanted to Know about Sex." *The Independent* (London), July 16, 2002.

INTERVIEWS

Nick Comfort, interviewed by Pagan Kennedy, 2007.

Robert Greacen, interviewed by Pagan Kennedy, 2007.

Marilyn Yalom, interviewed by Pagan Kennedy, 2007.

Genius on Two Dollars a Day

She's striding down a hallway at MIT with a bucket in one hand and a length of string in the other. Amy Smith—winner of inventing awards, and the brain behind such creations as the screenless hammer mill and the phase-change incubator—must fish some water out of the Charles River before she can teach her next class. As she lopes along, Smith describes the ordeals of testing water in remote villages. Her words spurt out. She's a woman on fast-forward, and she does not so much talk as download information. I'm jogging to keep up with this lanky, deeply tanned scientist and the bucket bumping against her leg.

We reach the massive front doors of MIT's main building and she pushes out into the crisp air and roar of traffic. She hurries toward the steps. Without interrupting her disquisition on water-testing, she perches sidesaddle on a banister. Elegantly poised, she slides down the handrail, still talking.

She lands on the sidewalk with a practiced leap. "You slide faster in the winter," she says, "when you've got a wool coat on."

If you're an inventor and you ride the banister below the pillared entrance of a university, there's a good chance that you could come off as way too cute. It's something Robin Williams would do if he were playing an inventor in a Disney movie. But Smith can get away with this kind of flourish. She is, after all, one of the brightest minds in a movement that sets out to prove that the best technology can be cheap and simple. The banister is a perfect example: it requires less energy than the stairs, and it's free.

In a culture where innovation means coming up with new functions for the cell phone or new spam-busting software, Smith is gloriously out of step. She designs medical equipment and labor-saving machines for people who live at the far end of a dirt road in Africa. Her inventions cost anywhere from a few hundred dollars to a few pennies. "You can't understand how important a grain mill is," she says, "until you've spent three hours pounding grain and gotten a cup-and-a-half of flour." It is this kind of understanding—of tedium, of tired muscles, of hunger pangs—that Smith brings to her designs.

***

An hour later, in Smith's "D-lab" class, students gather around a huge, black-topped slab of a table. It's the first semester of design lab, and these undergrads are learning about the politics of delivering technology to poor nations, how to speak a little Creole, and the nitty-gritty of mechanical engineering; during the mid-semester break, they'll travel to Haiti, Brazil or India. There, they will act as consultants in remote villages, helping locals to solve technical problems. Oh yes, and the students will also test village drinking water for dangerous bacteria.

Today, Smith is training them to do just that. Students practice, using a small pump to pull polluted water (from the Charles River) through a filter. Smith points to a piece of the water-testing rig—what looks like a silver barbell. "This test stand costs $600. Personally, I find that offensive." When the students work in the field, she says, they will be using a far cheaper setup—one that she patched together herself for about $20 last year, using a Playtex baby bottle. "You can do a lot more testing for the same amount of money."

Now the students have made cultures of Charles River water in petri dishes. The next step is to incubate the petri dishes for an entire day at a steady temperature. But how do you pull that off in a lean-to in Haiti, with no electricity for miles around? Again, Smith has a solution. She passes around a mesh bag of what appears to be white marbles. The "marbles" contain a chemical that, when heated, will stay at a steady 37 degrees Celsius for 24 hours. The balls are the crucial ingredient in one of Smith's inventions—

a phase-change incubator that requires no electricity. The design won Smith the 1999 BF Goodrich Collegiate Inventors' Award. The Centers for Disease Control and Prevention may soon endorse her incubator, and "from there it's not a big step to go to the Red Cross," she says. One day it could be a key piece of equipment at rural health clinics in Africa, Haiti, India—where doctors depend on intermittent electricity or none at all.

At the moment, however, Smith is still in the process of manufacturing the incubators—she has founded a startup company to handle the exigencies of getting her incubators up and running, and out into the field. "I have 6000 of these balls on their way here from China as we speak," she says.

Now, Smith wants to demo a bacteria test in the dark. She asks a student to cut the lights. He flips a switch. For a moment, nothing happens and then the room vibrates with a mechanical hum, and panels close over a bank of skylights in the ceiling. Everyone cranes their necks to watch. The panels move in menacing slo-mo, like something out of a James Bond movie. A few people giggle, as if they've suddenly become aware of the contradictions thrumming in this room—they've come to one of the best-funded technical institutes in the world to learn how to work with Playtex baby bottles.

A few weeks from now, Smith will give them one of their toughest lessons in the gaps between first world and third. The students will spend a week surviving on $2 a day in Cambridge—the equivalent of what the average Haitian earns.

Last year, Jamy Drouillard—who TA-ed Smith's class—performed the assignment along with his students. Drouillard grew up in Haiti, but that didn't give him any special edge. He laughs, remembering his chief mistake. "I bought a bunch of Ramen noodles, a bunch of spaghetti, and some ketchup. It got sickening after Day Three. Actually, before Day Three. I should have mixed and matched instead of buying ten boxes of spaghetti. In Haiti, people come up with creative ways of varying their food intake." He said the assignment drove home Smith's point quickly—living at subsistence level requires enormous creativity. The African farm-woman who finds a way to make a scrap of land yield enough cassava root for her family, she, too, is an inventor.

Last year, at a lavish banquet, Smith pulled out crackers from her pocket and nibbled while colleagues feasted around her—she was sticking to the $2-a-day assignment, in fellowship with her students. For Smith, the exercise must have been a snap. She had, after all, lived at the edge of the Kalahari Desert.

***

"I never got very good at the Bushman languages. My accent is really bad," Smith says. "There's fourteen different clicks and I always did the wrong ones. So I used to click as I walked around, trying to get it right."

In the late 1980s, as a Peace Corps volunteer, she was stationed in Ghanzi, a backwater of Botswana down a dirt path that could take as long as two days to travel. "Nobody wanted to live there. You got sent there for punishment if you did badly in a job."

Smith taught local kids math and science. She coached volleyball and ran the beekeeping club. To escape the stray cats that had taken over her bedroom, she slept outside her rooms, at the brink of the desert.

"I knew I wanted to stay in this country," she says, even though a certain loneliness was setting in—the loneliness of the nerd.

Smith's father taught semiconductor physics at MIT; her mother taught junior-high math. At the dinner table, the family would chit-chat about ways to prove the Pythagorean theorem. In her first couple of years in the Peace Corps, she missed the kind of people she'd known growing up in the orbit of MIT—people willing to engage, for instance, in passionate discussion about the innards of a motor. "I'd run into development workers who had no clue about engineering. They wouldn't understand that there was a way you could solve a problem" for the Africans. And she wanted very badly to solve the problems for the people she'd met, women ferrying water on their heads, grinding, washing, lifting, churning their lives away.

In 1987, she received word that her mother had died. She flew home to Lexington, Massachusetts. "When you're deep in grief it's harder to be tolerant of a society's excesses," she says. Wandering through a supermarket after the funeral, she marveled at the lunacy of her own country—an entire aisle just for soup? It seemed impossible to reconcile the two places she loved, to bridge the gap between America and Botswana.

About a year later, Smith was gazing out the window of her room, studying the expanse of the Kalahari Desert pocked by thorn bushes. Suddenly, she understood the arch of her life: she would learn how to be an engineer and bring her skills to a place like this. She sent away for applications to graduate programs. Fate— or rather, the kooky force that passes for fate in Smith's life— intervened. "A cat had kittens on my grad application for U Penn. It was just covered with placenta stains, and I didn't feel I could send it in. So that's how I ended up returning to MIT," she says. "That's how life always works for me."

***

Sometime after she got back, a professor suggested that she try to solve a problem that bedevils people who live in rural Africa. It involves the hammer mill, a no-frills, motorized grain mill that women use to grind sorghum or millet into flour. The hammer mill can do a job in just a few minutes that might otherwise take hours, which makes it a hotly coveted item in developing nations. But there's a built-in flaw: the mill uses a wire-mesh screen. When that screen breaks, it cannot easily be replaced, because parts like that are scarce in Africa and not easy to fabricate. So for lack of a wire screen, grain mills often end up in the corner of a room, gathering dust.

What was needed was something that could not only match the efficiency of the hammer mill but also use materials available to a blacksmith in Senegal.

A group of MIT students had come up with some ideas, but Smith, who had ground sorghum by hand in Botswana, knew they weren't fast enough. So she devised a system based on an elegantly simple element: air. She redesigned the machine to use the air passing through the mill to separate particles. The smaller ones—a.k.a. flour—get carried out while the larger ones stay behind. The resulting machine would cost a quarter of what its predecessors had and use far less energy.

But figuring out how to distribute her machine has turned out to be a far more vexing challenge. Originally, she worked with a group in Senegal—but they changed their mission before Smith had tweaked the design for the mill. Now, she's collaborating with a metalsmith in Haiti, who will translate the design specifications into French, which will allow her to bring the plans to an NGO in Mali that will aid local groups in manufacturing the machine.

For her screenless hammer mill, Smith became the first woman ever to win an MIT-Lemelson Student Prize. Past recipients of the high-profile award for inventing include David Levy, who patented not only the smallest keyboard in the world but also a surgical technique that speeds the splicing of severed arteries.

Smith has a rather tortured relationship with prizes. When I asked her about being named Peace Corps Volunteer of the Year while she was in Botswana—beating out 2,500 people—she offered up whimsical logic to explain why she didn't deserve it, something to do with a batch of brownies. Nor does she take her science awards too seriously. "Winning the Lemelson was helpful. Some people changed their attitude toward me after I won," she says, and leaves the rest hanging.

Smith, 41, has no kids, no car, no retirement plan, and no desire for a PhD. Her official title: instructor. Her life is like one of her inventions, portable and off-the-grid. "I'm doing exactly what I want to be doing. Why would I spend six years to get a PhD to be in the position I'm in now but with a title after my name? MIT loves that I'm doing this work. The support is there. So I don't worry." It was a good thing that she won the B.F. Goodrich award in 1999, she says, because back then she was stretching a three-month graduate-student stipend to last for a year, and didn't know how she'd pay her rent. The $7,500 came just in time.

Likewise, the inventors who most inspire her will never strike it rich. "There are geniuses in Africa, but they're not getting the press," she says. She gushes about Mohammed Bah Abba, a Nigerian teacher who came up with the pot-within-a-pot system. With nothing more than a big terra-cotta bowl, a little pot, some sand and water, Abba created a refrigerator—the rig uses evaporation rather than electricity to keep vegetables at 40 degrees.

Innovations that target the poorest of the poor are most effective when they have a certain no-duh quality about them—as soon as you hear an explanation of the way the thingamajig works, you can't believe that it took human beings so long to figure it out.

Smith, of course, aims to design such hidden-in-plain-sight tools and deliver them to the needy. But she also nurses a much grander ambition: to redefine invention itself. To this end, she has co-founded the IDEAS award at MIT; students work with a non-profit group to solve a problem for the disenfranchised. Last year's winners, for instance, included a team that put together a kit for detecting land mines, so that farmers in places like Zimbabwe no longer have to improvise with hoes and rakes when they tap the ground to see whether it might explode.

Success in the IDEAS competition, as well as in the kind of design that Smith pursues, requires humility, because your master-piece may end up looking like a bunch of rocks or a pile of sand. And, since you'll be required to do extensive fieldwork to under-stand the problem you're solving, it also demands the skills of a crack Peace Corps volunteer, someone who remains cheerful even when the truck breaks down, the food runs out, and you're the one who has to sleep next to the goat.

Unlike in other branches of engineering, women have the advantage here. "I know how to be self-deprecating. The traditional male engineer is not taught that way," Smith says. That engineer, were he trying to figure out an agricultural problem in Botswana, might consult with men—which wouldn't get him very far. "In Africa, the women are the farmers. Women invented domesticated crops. If you're talking to the right people, they should be a group of elderly women with their hair up in bandannas."

Unlikely as it may sound, Smith's brand of invention is moving into the mainstream. And that's because her clients—the disenfranchised in Africa, Haiti, Brazil, India—are increasingly able to secure loans.

***

In the late 1970s, Muhammad Yunus, an economics professor in India, tried an experiment: He loaned pocket change to poor villagers who lived near his university so that they could start small businesses. They did. And they paid him back. Grameen Bank grew out of that experiment, loaning to the rural poor in Bangladesh, and—contrary to conventional wisdom—making money. In the 1990s, microfinance caught fire. Thousands of institutions began extending loans to impoverished people in developing nations so that they could buy or lease the materials they needed to start small businesses.

Recently, commercial banks have been following suit—tailoring their services to poor people who live in remote villages, according to Elizabeth Littlefield, CEO of the Consultative Group to Assist the Poor (CGAP), a donor consortium on microfinance housed in the World Bank. Littlefield believes that the integration of microfinance into mainstream banking could bring billions of new consumers into the global marketplace over the next few decades. There have already been some surprising strides made. In India, for example, banks have set up solar-powered kiosks in out-of-the-way villages, giving clients access to financial services in places where there is not even electricity. But what will they invest in? The rural poor will need machines designed for their environment. And that will create demand for a new kind of technology.

***

In a barbeque pit near the MIT student center, pale blue smoke streams out of a trashcan and twists in the direction of the tennis courts. It smells of caramel. Shawn Frayne, a gangly guy with a shock of black hair and a skateboard tossed nearby, sticks a lighter down into the trashcan. He's trying to get a real fire going in the shredded sugarcane husks inside. He holds up one of his finished products— a piece of cooking charcoal that looks like a jet-black hamburger patty. It's made out of the parts of the sugarcane that aren't edible— that is, trash. These humble wads could help to solve a number of problems in Haiti: Poor people would be able to make their own charcoal rather than having to pay for the pre-fab variety; forests would no longer have to be cut down to make wood charcoal; and local entrepreneurs could use the recipe to set up small businesses.

Frayne graduated from MIT last year. He didn't like school much—except for Smith's design class, to which he is so devoted that he volunteered to put finishing touches on several inventions the class started last year. "I learned in an economics class that if someone has a good idea and they can implement it in a third-world country, they can dramatically change the economy of the country. I was surprised by how much technology can affect the well-being of a people. Amy showed me that someone's actually trying to make that kind of technology."

Smith herself stands by, trying to keep the wind from whipping her blond hair into her face. "We're working on a portfolio of designs like this charcoal that we can show to the Peace Corps or to NGOs, groups that are trying to help people start up small businesses," she says. In an era when rural people in Haiti increasingly have access to small loans to experiment with new machines and techniques, Smith hopes to supply the blueprints.

Frayne ducks down, pointing to the base of the trashcan. "If we were in Haiti, we'd use dirt to seal up the bottom of the can," he said. "But I couldn't find any dirt around here, so I used duct tape."

"In Cambridge, duct tape is the equivalent of dirt," Smith says, meaningfully. She loves duct tape and all it stands for. Last year, she taught a workshop on duct tape. She knows how to make a hammock and a kaleidoscope and full suit of armor out of duct tape. It's a very useful material, no doubt, but if she were on her two-dollar-a-day budget, she'd probably have to buy it on lay-away.

UPDATE:
The day after this story appeared in the *The New York Times* Magazine, Kofi Annan called Amy Smith and invited her to meet with him at the United Nations. A few weeks later, I heard from an administrator at the MacArthur Foundation. As a result of my story, the MacArthur people had decided to consider Amy for a "genius" award. She won. These days, she presides over the International Development Initiative at MIT.

Bird Brain

Call him Alex—just Alex. This celebrity goes by one name. Personal assistants swirl around his perch, offering him water, a massage, a toy. He eats only vegan food, of course, breakfasting on broccoli cooked for him by his personal chef. Japanese film crews want to shoot him. Visiting scholars want to watch him at work. Alex the parrot lives in a room at the Brandeis University Foster Biomedical Research Lab, where a team of grad students, a.k.a. "parrot slaves," cater to his every whim.

It's not his looks that got him here: an African Grey, Alex is covered in feathers the color of a sweatsuit. Rather, Alex has made it as far as a parrot can go in academia by proving himself to be exceptionally smart, or at least better educated than just about any bird on the planet. Dr. Irene Pepperberg, an adjunct professor of psychology at Brandeis, has been training him for 27 years, using a unique method that she calls the model/rival technique. Alex watches interactions between his trainer and another parrot or a human being, learns to answer questions correctly, and competes to show he knows the right answer—it's something like being a contestant on a parrot version of *Jeopardy.* And the technique works, dazzlingly. In the 1980s, Alex made headlines for being able to identify objects by shape, name and color with an 80 percent or better accuracy rate. Since then,

Pepperberg has managed to teach Alex the rudiments of spelling, addition, subtraction. Next year, she will publish a paper in which she argues that Alex can understand a simple form of the concept of "zero." Because of his sophisticated ability to communicate, Alex gives us a new window into the mind of a bird—in fact, Pepperberg is now working with another researcher to compare Alex's perform-ance to that of human children. His walnut-sized brain, which evolved along a different evolutionary path than our own, may turn out to function in remarkably similar ways.

I first came across Alex on the Web, when a friend called me over to her laptop and said, "You've got to see this." Together we watched a postage-stamp-sized video of the parrot. He cocked his head at a green key and a brown key. "Tell me what's different?" someone asked from off-screen. "Color," the parrot said in his little-girl voice.

After I saw his videotaped performance, Alex continued to perch in my brain and I couldn't seem to get him out. He was dis-turbing in the best sense of the word, one of those oddities that convinces you to expand your notions about what might be possi-ble in this world. But I must admit that I was also hooked by Alex's star power; in that video, he spoke in an impossibly sweet whisper that reminded me of the loyal (and fictional) flipper in *Day of the Dolphin,* who cooed "Fa loves pa" to its trainer and then swam off to its death; my eyes had welled up with tears even as I cursed myself for being manipulated by a stupid dolphin movie. What is it about talking animals that gets us right here?

In real life, Alex turns out to be nothing like the vulnerable naïf I had imagined. I find him squatted on a perch with his beak tucked into his neck, glaring at me with a half-hooded eye. Bits of feather, like aimless snowflakes, float in the air. Two other parrots—Griffin and Arthur, a.k.a. "Wart"—are mellowing out on their own perches. Dr. Pepperberg, a woman with a dramatic sweep of dark hair, sits in a desk chair nearby. She begins fussing with Griffin's perch, lowering it on its adjustable stand. "Alex has to sit as high or higher than the other birds," she explains. Of course—top billing.

"Want nut," Alex snaps.

Pepperberg decides this means he's ready to work. She arranges six blue, four red and two yellow blocks randomly on a tray. "What color two?" she asks, proffering the tray under his beak.

Alex bobs his head this way and that, gazing at the blocks suspiciously. Pepperberg repeats the question a couple of times. Alex seems to space out. Finally, he whispers, "Yellow."

"Good boy," she says.

Alex has performed feats such as this one correctly over and over again—allowing Pepperberg to argue that he can recognize quantities from one to six. But he has also thrown some hissy fits in between his star turns, and it was during one of these that he made his most recent breakthrough. When Alex refused to cooperate one day, Pepperberg decided to try something new. "OK, Alex, tell me, what color 5?" she asked, holding before him a tray with a set of two, three and six colored objects on it—but no set of five.

"None," Alex shot back.

Alex knew the word from other contexts. However, he had never been taught to find a word for an empty space, a zero. Pepperberg believes he improvised the answer. In a series of trials, Alex continued to use "none" over and over again, correctly, to mean zero.

Today, Pepperberg considers putting Alex through his "none" paces for me, but decides he's too cranky.

Alex, in fact, is winding himself up into a four-star, Hollywood huff.

"Want cork," he squawks. Pepperberg offers him a piece of the best damned cork a parrot could ask for—$75-a-bag sterilized primo that won't infect sensitive beaks. Alex chomps down on it, enjoying its chewing-gum-like texture. But in a matter of seconds, the cork goes flying across the room.

"Want cork!" he bleats. Pepperberg sighs and fetches it for him.

"Want cork! Want cork!"

Throughout the visit, Alex continues to spit out the cork and Pepperberg tirelessly retrieves it. "Hold it with your foot, Alex," she says at one point, but otherwise exhibits the profound patience of a woman who has been training parrots for decades.

In her forthcoming academic article—to be published in the *Journal of Comparative Psychology* in May—one paragraph reveals just how frustrating it must be to study parrot intelligence year in and year out. "Alex completely balked during testing for approximately two weeks," she reports. "He would, for example, stare at the ceiling, reply with a color or object label not on the tray, fixate on that label, and repeat it endlessly, interspersed with requests to return to his cage." Whether or not Alex understands nothingness is still up for question. But he clearly knows how to act like a star.

UPDATE:
Alex the parrot died in 2007, on the same day as the actress Jane Wyman, who had so deliciously played the villainess on *Falcon Crest.* I don't know why the two deaths immediately seemed to connect in my mind, but they did.

The Strongest Woman in the World

Cheryl Haworth, a teenager who has just become the world's top-ranked female weightlifter, straddles her croquet mallet. At 5 feet 10 inches and solid as a linebacker, she can heft more than 300 pounds, the equivalent of two refrigerators, over her head. But now it's high noon, and we're tapping wooden balls in one of Savannah's city parks, where a rococo fountain throws a chandelier of droplets into the air. Cheryl, who wears gym-rat attire, a t-shirt with a hole in it, seems to relish the odd figure she cuts among the Victorian-dainty wickets.

"Instead of going out to clubs, my friends and I play croquet," the 19-year-old tells me. Her best friend Ethan dangles his croquet mallet lethargically, his back curved in the S of a tall boy just out of high school. A huge crucifix thumps on his chest every time he moves. In the glare of Southern sun, the silver Jesus burns like a hood ornament. On the pocket of his plaid shirt, a button proclaims "God Bless Our Priests."

Cheryl and Ethan are riffing about the decorating of her new house. This evening, she will bid on a two-bedroom ranch on a suburban street, and if all goes well, they will move in together.

Because Ethan told me he favors a "swinging sixties" motif for the house, I had assumed the crucifix should be taken as an ironic gesture. It's a commentary, perhaps, on the pedophilia scandal in the Catholic church, the kind of fashion statement that brainy kids make at that age.

Actually, no. Cheryl's mother pulls me aside. She explains that under ordinary circumstances, she would worry about her teenaged daughter sharing a house with a boy. But honestly, Ethan? His career goal is to be Pope.

Ethan, in fact, plans to be a Catholic priest. But "there is a chance—it's probably miniscule and really unlikely—that I will one day be the bishop of Rome," he tells me. When I ask him how his decorating theme—hairy rugs, beads, Jetsons chairs—fits with papal ambitions, he dissolves into laughter. Cheryl answers for him. "Ethan is a bundle of contradictions." And then, she shoots him a look, "And when you're Pope, I won't be kissing no rings."

Cheryl, in any case, has her own grand ambitions, and they're a little closer at hand. This year, she set a record at the international weightlifting championships, giving her bragging rights as the world's strongest woman. But until she wins Olympic gold, no one will pay much attention. Cheryl took home a bronze medal in the 2000 Olympics. In 2004, if she is lucky, she will hunch in front of a barbell, throw it in the air and, perhaps, prove to television audiences and sponsors around the globe that she is really, truly the strongest woman.

***

She began with treehouses. Ten years old, she cut down pines and oaks and sliced them up. She dragged the logs into place and carried them up ladders. She reinforced the floors with T-shaped beams. When she'd finished one treehouse, she'd move onto the next. She tells me the story in a tumble, her two sisters chiming in, adding details about the wars they waged with neighbor kids. The Haworth girls know their way around this anecdote, as if they've told it many times, part of the way they explain themselves to outsiders. I picture the three of them as pint-sized frontierswomen, uprooting mighty redwoods, wrestling bears, bending railroad ties—their sense of themselves is so Paul Bunyan-esque that they seem to have emerged full-blown from a campfire tale.

Cheryl didn't become a strongwoman, officially, until age thirteen. Her father, wanting to hone her upper body strength for softball, drove her over to the Paul Anderson-Howard Cowen gym. The family had been reading newspaper stories about the gym for years. Its Team Savannah members, male and female, had swept the weightlifting championships.

With typical Haworth gusto, Bob marched into the gym and proclaimed that his daughter might want to try out for Team Savannah, though she'd never once lifted a weight. "You might have the strongest girl in the world here," he crowed. The women's coach led Cheryl away to test her mettle. "Next thing we knew the coach was whooping and a-hollering. 'This is the strongest girl I've ever seen.' There were 120 pounds on the bar."

Within weeks Cheryl began competing, cleaning up in the local meets. "I beat everybody in my first competition," she says, matter-of-factly. Soon she was jetting to international competitions. In 2000, she competed in the first-ever Olympics to include women's weightlifting as an official event. She, the new-comer, won the bronze medal. Reporters seemed less impressed with Cheryl's feat than they were stunned by her size—over 300 pounds.

In front of banks of tape recorders, they hit her with a barrage of questions that would have driven most other teenage girls into hiding. Had she ever been on a date? What exactly did she eat? Didn't she feel self-conscious about her body? Cheryl answered with aplomb. "I'm not trying to be small. I'm trying to be strong," she reminded them.

Whether she wanted to or not, her very existence issued some kind of challenge to national assumptions about the good life. "There's something in us that wonders if she can be truly happy," a *Boston Globe* columnist declared.

According to her agent, George Wallach, Cheryl stands to make a lot of money if she can prove she's happy. "She's a big gal and for a lot of women who aren't completely comfortable with themselves, she makes them feel better." Wallach envisions numerous ways of cashing in on this—a plus-sized clothing line, motivational speaking tours, maybe even "an animated series, 'Cheryl Haworth, Strongest Woman in the World, Saves the…whatever.'"

***

Cheryl cannot afford to think about her body as a symbol. She has to think about the Chinese. They own women's weightlifting. She must stay big enough to take them on. "Cheryl can weigh whatever she wants to, provided that she's got her speed and strength," according to her coach, Michael Cohen.

What about the long-term health effects of all this bulking up? "Look at gymnastics," he says. "You look at those girls and you think, 'My god, that's child abuse.'" He pauses for effect. "No, that's not child abuse. That's athletics."

To compete in the Olympics you cannot be romantic about your body—you will need to starve it or muscle it up. You must endure being measured, analyzed, and poked. This becomes clear to me when Cheryl's drug tester shows up. It's evening. I'm slumped over the Haworth's kitchen table, exhausted from a day of trying to match their pioneer spirit. A thirty-ish woman in a linen blouse lets herself in the door; a registered nurse, she makes surprise visits to the house once a month. From the moment she arrives until Cheryl produces enough urine for her to analyze—which is a lot, I gather—the nurse cannot let Cheryl out of her sight. She shadows Cheryl for three hours, jumping from her chair whenever the teenager pads into another room.

"She's gotta watch," Cheryl says, of the actual urine-collection process. "I have to put my shirt up to here and my pants around my ankles," she adds and sighs. She takes another gulp of Coke, trying to fill her bladder, to get the ordeal over with.

***

Cheryl is cruising the highway with the AC up and Dave Matthews blasting on the stereo, working a toothpick in her mouth, lounging in the cushions of her Impala SS's driver seat. The black, lowrider-ish car reminds me of a shark. She bought it with cash. When you're inside it, surveying Savannah through the tinted windows, the trees pop out in surreal blues and greens. The clouds all have silver linings.

"It's odd, because I'm a woman and I'm a weightlifter so I should be all about feminism," she continues. Then she glances over at me, eyebrow arched in question. "Is that the word— feminism?" Recently, NOW honored her with one of its Women of Courage awards. "I didn't do anything courageous," Cheryl scoffs. "I lifted weights."

I can see why feminism might seem unnecessary within the Haworth family. The eldest sister, Beth, "was always MVP in every-thing," according to her father. Sixteen-year-old Katie has placed twice in national weightlifting championships, but says she won't let athletics mess with her law career. "What do you want to be?" I ask her. "Supreme Court Justice," she answers, without missing a beat. Cheryl manages to keep up her grades at the prestigious Savannah College of Art and Design while training for the next

Olympics. She has shown me her drawings, photo-realistic portraits in black and white, the muscles of her subjects lovingly rendered, the eyes wet with life. The Haworth girls take it for granted that they will rule the world.

Bob Haworth tells me how he and his wife, Sheila, skimped on sleep to shuttle their daughters from swim practice to mock law trials to music lessons. He's smaller than Cheryl, but I can see echoes of her in the way he walks, the side-to-side swagger of a guy who once wrestled on a college team. His folksy voice would be perfect for narrating a Christmas special.

Sheila has the rolled-up-sleeves manner of a woman who has seen her share of split-open skulls. "We get the traumas, the gunshot wounds, kids riding their bikes with no helmets on, the adults riding the four-wheeler getting paralyzed from it flipping it over," she says about her work as a nurse on a team that specializes in head, neck, and back injuries. "We see the results of a lot of stupidity," she adds.

She regards sexism as just another brand of stupidity. "My dad was a brick mason," she says. "I had a cousin Walter, who was the same age. Walter got to go work with his dad. My dad wouldn't take me. I could do just as good a job as Walter. If my dad had taken me, I would have been a brick mason."

"Is she talking about the bricks?" Cheryl says.

"It makes her so mad," Katie says.

Sheila and I exchange a look. Her girls don't get it, and she's proud they don't. The Haworth parents have taught their daughters to expect big sky, a wide-open future where they can ride like cowboys into any sunset they choose. The Haworths have a certain genius for optimism. But when I ask Cheryl how she became so confident, she does not immediately think of her family. She credits Savannah. "In the South, people don't care. It's different from California. I can wear the same shorts five days in a row. Being a woman and a weightlifter is more cool to people than odd." Her fans—mostly middle-aged men—salute her everywhere we go. "Hey Champ," they call. Or "strongest woman!" Or "Saw you in the paper!" In a restaurant, one beefy guy pulls her aside to consult about training tactics. An old man calls her over, and his wife says, "You're even prettier in real life than in your pictures."

***

I had expected chrome and flashy mirrors, a downtown location, maybe an underground parking garage with spaces reserved for the coaches. After all, the Anderson-Cohen gym sent more Olympic weightlifters to the 2000 games than did the official Olympic facility in Colorado. But the gym is unpretentious. It hunkers at the end of an access road that I had trouble finding on the map. The athletes park on the grass.

As soon as I swing open the door, the din hits me, the THUNK-thunk-thunk of barbells falling from six or seven feet in the air and then bouncing across the plywood platforms. Lifters sprawl on plastic chairs, waiting their turn—gawky kids training for basketball teams, tattooed guys, women who could pass for action figures, heavy kids who could pass for bookworms.

I find coach Cohen squatting under a barbell. When I get him in a chair, he's still squatting, elbows balanced on knees, as if he's ready to pounce, a man made entirely out of fast-twitch muscles.

Cheryl, he says, just returned from the Junior Championships, where she successfully defended her title as strongest under-20-year-old girl in the world. She needs rest. But in a few weeks, "we will bust her behind. She'll be here all day."

What about the Haworth family? I ask. How does he think the parents managed to raise three such confident young women?

Cohen doesn't seem to be interested in that question. What interests him is that each of the Haworth girls fall into a different weight class. "You've got Beth who weighs 130, 125 pounds. You've got Katie who weighs 170 and you've got Cheryl who weighs over 300."

Just then, two of the three Haworth weight classes appear— Cheryl and Katie. They announce their mission for the afternoon: ear-wax candles. Supposedly, you can stick one of these folk remedies into your ear, light it, and it will suck all the gunk out. They plan to find out if this is true.

But first, the girls take over the gym office. Katie swivels back and forth in the chair behind the desk, answering the Anderson-Cohen phone when it rings. I pull out my tape recorder, intending to interview her about her future, but it's just too hot. Arrows of sun scorch the desk. Katie and I become absorbed in winding weightlifting tape around the office pens. Days later, when I have returned home, I will find one of these swaddled pens in my bag.

Cheryl's cell phone bleeps. It's Ethan. Hand it over, I tell her, because I realize that a certain question has been gnawing at me. "Ethan," I say, "What are you going to do after you become Pope? What's your plan for the church?"

He admits he doesn't have one. Cheryl grabs the phone away. "You don't have a plan? Ethan, you've got to have a plan!"

***

When we head over to the house she wants to buy, two real estate agents perch on a sofa. Cheryl sprawls on the floor. The sweet little two-bedroom ranch with a grill on the back porch requires a down-payment that's a bit beyond her means. Now Cheryl pitches mortgage terms at the agents. They treat her with the deference due a celebrity who may be trading up to a four-bedroom-with-swimming-pool in a few years.

"I don't think about it at all," Cheryl told me earlier, about body size. However, if she follows her agent's advice, she will have to. He wants her to cash in on her body's power to explode assumptions about female strength, and the challenge she poses, willingly or not, to one-size-fits-all culture. It's not the easiest route to fame, but it sounds like a plan.

UPDATE:
At the 2004 Athens Olympics, Cheryl Haworth tore a ligament in her elbow. Once the favorite for a Gold, she came in sixth. Two years later, she was arrested for drunk driving. After an argument with her coach Mike Cohen, she decided to leave his gym. She sold her house and moved to a dorm in the Olympic training center in Colorado. I have no idea what happened to Ethan.

Battery-Powered Brain

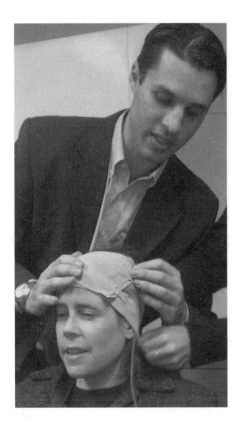

This spring, Stuart Gromley hunched over a desk in his bedroom, groping along the skin of his own forehead, trying to figure out where to glue the electrodes. The wires from those electrodes led to a Radio Shack Electronics Learning Lab, a toy covered with knobs, switches and meters. "For ages 10 and up," the instructions on the box recommended. Gromley, a 39-year-old network administrator in San Francisco, had bought a kiddie lab because it had been decades since he'd last tinkered with electricity; he hoped its instruction sheets would help him cheat his way through the experiment he was about to set up. He couldn't afford to make mistakes. He was about to send the current from a 9-volt battery into his own brain.

His homemade machine was modeled on the devices used in some of the top research centers around the world. Called transcranial direct current stimulation (tDCS), the technology works on the principle that even the weak electrical signals generated by a small battery can penetrate the skull and affect hot-button areas on the outer surface of the brain. In the past few years, research papers have touted tDCS as a non-invasive and safe way to rejigger our thoughts and feelings, and possibly to treat a variety of mental disorders. Most provocatively, researchers at the National Institutes of Health have shown that running a small jolt of electricity through the forehead can enhance the verbal abilities of healthy people. That is, tDCS might do more than just alleviate symptoms of disease. It might help make its users a little bit smarter.

Say "electricity" and "brain" in the same sentence, and most of us flash on certain scenes from *One Flew Over the Cuckoo's Nest*. But tDCS has little in common with shock therapy. The amount of current that a 9-volt battery can produce is tiny, and most of it gets blocked by the skull anyway; what little current does go into brain tissue tends to stay close to the electrodes. By placing these electrodes on the forehead or the side of the head, researchers can pinpoint specific regions of the brain that they'd like to amp up or damp down.

Some researchers believe that if tDCS continues to pan out, a consumer version of the machine might someday appear on the market. The final product might look like an iPod attached to a hat with electrodes in its brim—available with a prescription from a doctor. Still, it's very simplicity might prove its undoing. How would a medical-supplies company make money off of a gizmo so rudimentary that it sounds like a 7th-grade science fair project? The cost of parts—electrodes, a battery, a resistor—could be as low as $10.

In fact, a small clique of hobbyists on the Web have already begun to discuss how to make the machines at home and where to put the electrodes. Like ham radio operators of the brain, they share advice with their fellow tinkerers. "I accidentally found a way to make GREY FLASHES IN MY VISION using a 9v battery. Don't you try it," says one hacker in an online forum. "Here's how not to do it," he adds, and then provides instructions.

Gromley is one such hobbyist. He says that the first time he read about tDCS machines, he immediately decided, "I'm going to build one of these." He'd been suffering from bouts of depression since he was a teenager; antidepressant medication only made him feel worse; now, he hankered to find out if tDCS might give some relief. And he was curious about a technology that might let him fiddle with the knobs of his own personality, to experience something that might alter his consciousness. "I'm always thinking, 'What would it be like to be another person?'"

And so he set up the Radio Shack kit on a paint-splattered TV tray—the "art space" in his bedroom. He had bought sponge electrodes off of eBay, and he made an educated guess about how to position the electrodes—one on a temple area and one on the brow—based on the medical studies he'd pored over. When he flipped on a switch, current ran from the battery through a resistor and then into wires and into his prefrontal cortex. He leaned back in his chair with his eyes closed, wondering if he felt anything. That's when he saw the flash—what he describes as a "horizontal lightning bolt"—that seemed to arc from one side of his forehead to the other.

"No, that didn't happen," he thought to himself, and tried to calm himself. Then, a few minutes later, he shifted in his seat, the wires jiggled, and he saw lightning again. Gromley yanked off the electrodes. He began searching on Google, using keywords like "tDCS" and "flash" until he found a study that reassured him: those spots of light were harmless.

So Gromley went back to his experiment. Flash, flash, flash. He rearranged the electrodes several times before he found the sweet spot.

"Do you remember the first time you drank coffee? It was like, 'Oh my god, if I'd known how good this was, I'd be drinking coffee all the time.' Well, [tDCS] wasn't exactly like the first cup of coffee," Gromley says. "It was more like a cup you might have in the first month of drinking coffee. It's like, 'Hmm, I don't feel bad. I feel alert. I feel up.'" Actual coffee had long ago ceased to pull

Gromley out of his depression, but now he had the electronic kind. He used the machine about once a day for a month, and then he found that he no longer needed it much. Lately, he's been spending long hours at work—too busy to be depressed.

Of course, one man's Radio Shack adventure does not a study make, and the effects that Gromley felt could as easily be attributed to the placebo effect as to a tickle of electricity. Needless to say, the researchers I talked to cautioned against trying this sort of thing at home, although they had a grudging respect for anyone with the pluck to do it. "In the past, a lot of scientific discoveries were made by amateurs who experimented on themselves," according to Peter Bulow, a psychiatrist at Columbia University. He said that a recent safety study found that tDCS causes no damage to brain tissue, but cautioned that any cutting-edge treatment comes with unknown risks. Bulow himself has just submitted a proposal to study the effects of tDCS on 20 depressed patients.

He's in good company. Teams of researchers are experimenting with battery-powered electrodes at the National Institutes of Health, the Harvard Center for Noninvasive Brain Stimulation, and at the University of Göttingen in Germany, among other centers. They're exploring tDCS as a treatment for depression, chronic pain, addiction to cigarettes, Parkinson's disease, as well as motor disorders caused by stroke and neurodegenerative diseases. The gizmo, still only a few years old in its present incarnation, has become the Ronco Brain-O-Matic of the research world: a device that promises endless uses. But it remains to be seen whether it will prove itself as a truly effective therapy for any one disease.

***

In 1962, a 30-year-old woman shuffled around a hospital in England with a battery pinned to her dress. Two silver electrodes, wrapped in gauze, winked above her brow, like a second set of eyes. She'd spent half her life in mental asylums: when she was a girl, her father had shot himself, and afterward she'd become convinced that other people could see a mark upon her. Nothing had lifted her malaise: not even shock treatment. When she arrived at Summersdale Hospital, she muttered "Get rid of me" in response to questions. Researchers (J.W.T Redfearn and O.C.J. Lippold) attached the two positively charged electrodes to her forehead, with the cathode on her knee, to see whether they could use battery power to ease her out of her depression. They kept her brow area bathed in electricity for as many as eleven hours a day, three treatments a week. She began to sleep soundly, no longer tormented by nightmares; she ate well; she prettied herself up; she found a boyfriend. She became, the researchers said, a "different person."

That year, she was just one of several dozen people wandering the halls of Summersdale Hospital with electrodes plastered to their foreheads and batteries on their lapels like boutonnieres. In an earlier study, Lippold and Redfearn had found that they could change the personalities of their subjects with electrical stimulation: positively charged electrodes on the forehead caused people to giggle and chat. Under the influence of negatively charged electrodes, people shut down, became silent and apathetic. Some of the patients had so enjoyed the positive electrodes that they asked for the "battery treatment" again. And so Lippold and Redfearn launched this new study; this time they would expose people—many of them severely depressed—to long sessions of electrical stimulation. The patients were allowed to go home with electrodes glued to their heads, the battery still buzzing. Almost half of them experienced miraculous recoveries. A shell-shocked World War II veteran compared the effects to a snoot of a whiskey—"I feel quite all right," he crowed, after he'd been stimulated.

In the decade that followed, other researchers tried to replicate these effects. They produced inconsistent results. Nowadays it's clear why: researchers applied currents that were too small and glued electrodes to the wrong parts of the scalp. "They used some parameters of stimulation that we know now are not effective. They didn't have the information that we have now," according to Felipe Fregni, an instructor in Neurology at Harvard Medical School. Because the battery-powered electrodes seemed to be unreliable, the medical community lost interest in brain polarization.

Then, in the 1980s, researchers found a much more powerful way to stimulate isolated buttons of the brain. Called repetitive transcranial magnetic stimulation (rTMS), the technique uses electromagnetic radiation—which can easily pass through the skull—to create localized electrical fields near the surface of the brain. The effects of rTMS are dramatic and reproducible. Place the machine's wand on one part of the scalp and the patient will lose her ability to talk; move the wand to another spot and her leg will jerk.

In essence, rTMS is a souped-up version of the old battery-and-electrodes treatment. But it also comes with greater risks: it can trigger seizures, an arm that won't stop shaking or a patient who slumps over in his chair. However, such side effects are rare, and rTMS has proven to be a powerful tool for mapping the brain and modulating brain activity; in the 1990s, researchers began using it (along with other new technologies) to draw blueprints of neural function. This new knowledge, in turn, meant that the battery treatment became relevant again—because now scientists could design a more effective tDCS machine. In the late 1990s, a team at the University of Göttingen enlarged the electrodes and covered them in sponges in order to allow more current to pass through the skull. They also created new protocols for placement of the electrodes, aiming them with greater accuracy at hotspots such as the motor cortex. Just a few years ago, these design changes began to pay off in studies that showed promising results: the new, improved battery treatment could quicken the tongue and the hand. It could make people smarter and faster, if only by a small margin. When German researchers trained the positive electrode on the motor cortex, their human subjects became significantly faster at learning to hit a keyboard in response to a visual cue.

Other researchers confirmed this provocative finding: brain stimulation could enhance performance in healthy people. For instance, a 2005 study from the National Institute of Neurological

Disorders and Stroke (NINDS, an affiliate of NIH) found that tDCS stimulation revved up people's verbal abilities. They were able to generate longer lists of words starting with, for instance, F or W within a time limit. This has implications for victims of stroke and other neurodegenerative conditions—if tDCS can enhance the performance of healthy people, perhaps a machine could help pull lost words and hand movements out of damaged brains. For some patients, a wearable brain machine represents one of the few, dim hopes for recovery.

***

Georg Gabriel leans back in an office chair, so that it makes little creaking sounds underneath him. We're in a narrow room in an NIH complex. A lab assistant is standing behind Gabriel, running a tape measure around his scalp and carefully parting tufts of his white hair to mark him up with a Sharpie pen. Gabriel, who has been measured and Sharpied a lot these days, barely notices. "They told me that my life expectancy with this affliction was about five years," he tells me, with disarming good cheer. He's pink with apparent health, this 78-year-old man who until recently swam several miles a week. This morning, for his last session of tDCS testing, he's dressed in business casual: button-down shirt, "Nantucket red" slacks faded to a soft pink and boat shoes. He looks entirely put together. What you can't see is his brain: the nerve cells are dying off throughout the cortex; and the parietal lobe—that switching-house of sensation—may have already shrunk down. Gabriel has a rare condition called corticobasal syndrome, a degeneration of brain tissue with symptoms that often mimic Parkinson's disease. His movements are slow and dreamy. Earlier this morning, the lab worker put him through a battery of tests to rate his motor skills. On a finger-tapping test, Gabriel punched at a lever with such labored movements that I found myself leaning forward in my seat, willing him on.

Now the lab assistant glues one sponge electrode just above and behind Gabriel's left ear and another above his right eye; Gabriel is about to perform all the tests again, this time under the influence of tDCS stimulation. While the lab worker winds tape around his head, Gabriel tells me that the NIH researchers have asked him for permission to do an autopsy on his brain. He remarks, crossing one leg over the other casually, that he's inclined to give it to them. "That's a 'no brainer' decision," he quips, and then chortles at his own joke.

When the tDCS machine is on, Gabriel says he can't feel it at all. No tingling. No nothing. Neither Gabriel nor I know whether this is a "sham" stimulation or real. The electrodes might be attached to some area of the scalp where they would have little effect on motor function, or they could be aimed at prime real estate in the brain, one of the spots that the researchers hope will respond to exactly this kind of stimulation.

During most of the testing—Gabriel has to kiss the air, pretend to vacuum and wave goodbye—he continues to move in slo-mo. But on the finger-tapping test, he suddenly seems to gather himself. He looks as if he's been put on fast-forward, his hand jerking so fast that it doesn't seem part of him. His high score without the electrodes was 49; now, electrified, he fires off 66 taps. Even the lab assistant blinks with surprise.

After the testing is over, we learn that Gabriel was in fact receiving real rather than sham stimulation. Today, the positive electrode was placed over the area of the scalp that corresponds to the parietal lobe. However, until the data is compiled for all the patients in the study—and further studies that will surely follow this one—it's impossible to say whether DC stimulation can in fact enhance the plasticity of damaged brains.

"All of our good results to date have been in healthy subjects. I haven't seen convincing evidence that you can do much to help a brain that is badly damaged. It may be that there's no point in trying to polarize busted tissue," according to Eric Wassermann, the chief of brain stimulation in the Office of the Clinical Director at NIH's NINDS and one of the designers of this study. Despite inconsistent results so far, he and other researchers continue to explore DC stimulation for patients with widespread brain degeneration.

Such patients have been bypassed by recent advances that have helped, for instance, sufferers of Parkinson's Disease. In the past few years, surgeons have begun to use a technique called deep brain stimulation (DBS) to quiet the tremors and stiff gait that become debilitating during a Parkinsonian decline. After drilling small holes in the scalp, the surgeon threads wires deep into the brain to implant a chip near the cluster of cells that is sending out errant signals. For Gabriel, such a focal intervention would not work. In cases such as his, where disease sprawls across a lobe, tDCS could offer an edge. And, too, a cheapo electrical thinking cap, if it works, would offer a huge advantage over other stimulation techniques: No drills bore through the skull. No wires snake through brain tissue. No pacemaker-like machines get implanted under the skin.

I ask Wassermann what the tDCS machine might look like, if it ever hit the market—would it resemble an iPod?

"The brain-pod!" Wassermann jokes. "It should play music, receive calls and…shoot like a gun." Then he grows serious. "It could be very simple and wearable."

Wassermann believes that if we're ever to have a Brain-Pod in the United States, it would likely be tested and developed by the military first. But he doesn't rule out the chance that a private company would bankroll tDCS, if it continues to perform in the lab. "It is unlikely that any [company] would do this unless they were guaranteed a market share, and the only way they could be guaranteed a market share would be if they had a patent on some important part of the process. I think we know so little about it at this point that there may be patent-able parts."

However, Wassermann is not eager to put this device into the hands of consumers; he's concerned about the ethical problems it poses. "I would not be in favor of this being an iPod. Not yet. Not until the issues of safety and fairness have been resolved."

A while ago, someone suggested to Wassermann that he take some tDCS machines to a nearby university and wire up half the students in a classroom before they took a test. Would the battery-powered kids do better? "I thought the ethics of that sort of application were questionable because you don't want to advantage people who can afford something that others can't," Wassermann says.

Of course, these machines could be as cheap as clock radios or coffee makers. So arguing about the ethics of Brain-Pods might be an exercise in futility; if tDCS turns out to produce strong effects, the machines will pop up everywhere, whether we like it or not. "It's an interesting phenomenon, if this were an effective treatment, to have it get completely loose," Wassermann says. "I'm not excited enough about [tDCS] as a panacea or a great social evil at this point to be very worried. But if it were very potent, it will be all over the place. The Chinese would flood the market with gizmos. This could get completely out of control. It could be like blogging. Everybody could be a brain manipulator."

***

In October, a group of researchers gathered around a conference table at the Harvard Center for Noninvasive Brain Stimulation. Fregni, his hair slicked to the side in the manner of a 1920s tycoon, wielded a remote control, flashing images onto a white board. About twenty-five scientists—from Thailand, Brazil, Bolivia,

Israel, and Germany, among other countries—crowded the room. Most of them knew little about tDCS, and so Fregni was delivering an introductory lecture, what might have been called Brain Zapping 101. The graphs Fregni projected on the wall created a frisson of excitement. The audience oohed and aahed. They lifted digital cameras and snapped photos. You could feel it—the buzz that this technology is beginning to generate among the clique of researchers enchanted by both brains and gadgets.

At the end of his lecture, Fregni announced he would demonstrate tDCS. Did anyone in the audience want to try it out? Silence. The scientists gazed around, waiting for someone else to shoot a hand into the air. And then the room erupted into laughter at the collective reluctance to be wired up.

Before I quite realized what I was doing, I heard myself say, "I'll do it." My hand waved in the air, seemingly of its own accord. It was one of those moments when your body reacts while you're brain lags a second behind. My heart seemed to beat everywhere, my hands, my feet, my face. Why hadn't anyone else—any of the experts—volunteered? Now, I was teetering toward the front of the room. Shirley Fecteau, another Harvard researcher, guided me to a chair.

She and Fregni placed the sponge-covered electrodes on the top of my head, in the two spots where I might grow bunny ears, if I were a character in a fairy tale. This position, which targets the prefrontal cortex, is used to treat depressed patients. Someone wrapped an Ace bandage around my head so tightly that I began to feel headachy. I have lots of hair, and so the bandage began to slide upward. Someone pushed it back in place and I could feel fingers on my scalp, checking the position of the electrodes. Clearly, the Ace bandage alone wouldn't do the job. So Fecteau found a giant elastic band and stretched it vertically around my head so it cut into my cheeks. For the rest of the experiment, it squashed my windpipe, like an especially tight strap of a birthday-party hat.

Fregni showed the control box to the audience, a black brick with a meter and a few knobs on its face. The wire from that box dangled along my arm and went up beyond the line of my vision— to my head. That's when it hit me: They really were going to send electricity through my skull. Fregni turned the switch. The sponge on the left side of my scalp began to prickle, the way poison ivy will after you scratch it. The elastic band made me gasp for breath. The Ace bandage strangled my forehead. The room flashed as members of the audience took photos, and I tried not think about how

I must look with all the elastic pinching my face and my hair stick-ing every which way. Rather than an elevated mood—which the treatment was supposed to bring on—I felt mortified to be on display in mental-patient drag.

Fregni kept me hooked up for only five minutes, long enough to demonstrate the equipment but not to have much of a clinical effect. Then he freed me. I shuffled back to my chair, still trying to smooth my wet hair back into place. And now, as if by delayed reaction, euphoria overwhelmed me. I felt all fluttery, as if I'd just stepped off a roller coaster. Maybe the electrodes gave me the high. Or maybe I was just elated to leave the stage. It's hard to say.

In the midst of my intoxication, a thought came to me: I've touched my own brain. Before this moment, I had always thought of my brain as imperious and remote, like a queen who issued commands from a red-velvet room high up in a tower. "Worry ceaselessly!" my brain might decree, and I would have no choice but to obey. But now, I had tried to turn the tables. I had sent my prefrontal cortex a command made out of electricity. "Cheer up!" I'd ordered. And now, maybe, just maybe, it had heard me.

Vermin Supreme Wants to Be Your Tyrant

Vermin Supreme—a 43-year-old activist and street-theater performer—swaggers toward Faneuil Hall to take on the Democrat groupies. Inside the building tonight, nine candidates will debate each other live on CNN. Vermin Supreme plans to stay outside, where the TV trucks splash spotlights onto the cobblestones. A tribe of John Kerry people wave blue signs and scream in unison: "Ker-REEE, Ker-REEE." Many of them wear that mob-zombie expression on their faces—the glassy look of people who have been yelling one word for so long that it has turned into nonsense.

Vermin Supreme pushes his way toward the Kerry-ites. A few of them have to hop backward in order to avoid the pointy wingtips of the eagle lashed to his torso. He hoists his megaphone—which confers upon him the electronic voice of authority. "Where does John Kerry stand on mandatory tooth brushing?" he demands. "Is he soft on plaque?" A few college kids break off to listen to the tirade. You can see it in their faces; suddenly, they're no longer members of the Kerry gang. They're just their ordinary selves again, exchanging glances with each other. Who is this guy?

As he passes through the crowd, Vermin Supreme spreads that kind of puzzlement wherever he goes. He has spent years figuring out how to transform a group-thinking throng back into a bunch of individuals. This is his art form.

"Vote for me," he tells a gray-haired lady, peering at her from under the rubber boot stuck on his head, the foot end of which points up at the sky. "I'm running for...something." In fact, in the past fifteen years, he has set his sights on a number of political offices, all of them fictional: Tyrant, Mayor of the United States, Emperor of the New Millennium.

He seems at first to be a flamboyant-yet-sane hippie who is making a point about civil rights. With his wife, Becky (who asks that her last name be withheld), he roams the country to present his own brand of performance art at anti-war rallies and Republican pancake breakfasts. The couple fund their peripatetic activism by paying as little rent as possible and working odd jobs. But it's not as simple as that. When he goes home—to a shack in the woods of Massachusetts—he's still named Vermin Supreme. He's Vermin Supreme twenty-four hours a day, every day, no vacations.

THE PONY

Last year, I ran into Vermin Supreme at an anti-war march. Instead of the Visigoth costume, he had shown up in his Weirdo Lite ensemble—a Satan mask, megaphone, and sensible shoes. His job that day, as he saw it, was to boost the morale of the marchers, to be a sort of Bob Hope of the revolutionary army.

People swarmed around us, chanting, beating drums. Some guy screamed, "What do we want?"

"Peace," the crowd answered.

"What do we want?" the guy screamed again.

"Peace!" Now the river of people roared the word. The sound boomed through my chest. No one was laughing.

"What do we want?" the guy demanded again.

And this time Vermin Supreme pointed his megaphone at the sky. "A pony!" he screamed, his amplified voice rising over the roar.

Next time around, pretty much everyone in the crowd had defected to Vermin's chant. "What do we want?" "A PONY!" hundreds of people hooted. Some young women near me bobbed up and down. "A pony, a pony," they squealed.

Vermin Supreme has spent years working for peace, but what he really wants is a pony. He wants cotton candy and a funhouse mirror. He wants to topple the politicians from their pedestals and replace them with plastic chickens. He wants us all to live in a constant state of participatory democracy.

We're bound to disappoint him.

THE NAME

It's legal. "SUPREME, Vermin Love," reads the government-issue font on his drivers' license. He took the name in 1986, when he was booking bands for a grungy rock club. "All booking agents are vermin, right? So I decided to be the most supreme vermin. I schmoozed people in character as Vermin Supreme, wearing a tacky suit and chomping a cigar." When the job ended, he could not let go of the character. He decided to run for mayor of Baltimore in his plaid leisure suit. If all politicians are vermin, he reasoned, why shouldn't citizens vote for the best vermin available?

And Vermin Supreme—the name as well as the passion for cartoonish gestures—began to leak into his private life. His wife has been calling him Vermin for so long that the word rolls off her tongue like any other endearment, honey or darling or sweetie. At this point, even his mom calls him Vermin.

THE BOOT

Often, when Vermin Supreme shows up at a political event, it's the boot that causes trouble. Security guards confer with each other through their walkie-talkies and decide that the rubber galosh on his head has to go.

"What's the problem with the boot? Why is it so subversive?" Vermin Supreme wants to know, though the answers are obvious. The boot makes him the tallest man in the room. The boot gathers an audience around him. It draws cameras and microphones.

"That boot is like Wonder Woman's tiara," according to Darren Garnick, the producer of two PBS documentaries about fringe candidates. "If Vermin hadn't worn that boot on his head, I never would have noticed him. There were plenty of other guys who share the same values, but they don't have boots on their head, so no one listens. Maybe the boot is an indictment of the media."

Indeed. We journalists will follow a guy with a boot the way a trout will go after a shiny plastic worm. The boot promises a good story. Vermin Supreme knows this. He has packaged himself as a made-to-order wacky sidebar for newspapers to run during campaign season. Like the politicians that he mocks, Vermin Supreme presents a version of himself, and it's nearly impossible to see the real guy underneath.

When I asked to follow him around on a "typical day"—that is, to watch him cope with the exigencies of being Vermin Supreme on the job and at the supermarket—he only laughed. He made it clear that if he had any typical days, he would not offer them up for inspection. He requested that I keep his hometown a secret. And he refused to specify what kind of blue-collar job pays his bills, although I know, because I've snooped around, that he has worked construction in the past.

But it wasn't just his edicts that kept me from learning more about the private Vermin. During several hours of interviewing, he regaled me with anecdotes that had a pre-packaged quality, as if he'd told them many times. When I probed for deeper insight— Why does he use the name 'Vermin' in his private life? When did he feel his first twinge of political consciousness?—his words simply ran out. He didn't have answers about his own motives. He didn't seem to know what made him tick. In that way, too, he reminded me of a career politician. His own inner life bores him. He's interested only in his public self.

PEACE

Back in 1986, Vermin Supreme lived in Baltimore, a down-and-out art boy. One day, he heard that the Great Peace March for Global Nuclear Disarmament would pass through town, the cross-country parade of walkers opposed to military buildup. "I went to Memorial Stadium to check it out, and I was floored," Vermin Supreme remembers. He surveyed a parking lot that had transformed, overnight, into a town. "It was an amazing mobile community. They had Porta-potty trucks. They had kitchen trucks. They even had schools for the kids. It was so impressive."

A week before, in Two Greeks Restaurant, he had announced to his friends a new performance-art venture: He would run for mayor of Baltimore as Vermin Supreme. Now, as he surveyed that parking lot, his prank took on deeper meaning. He would align it, somehow, with this massive, mobile outcry for peace.

"I went to the thrift store and bought a sleeping bag and t-shirt and I joined that march," he said. And after that, he signed up with Seeds of Peace, a group that conveyed around the country to furnish demonstrators with food and sanitation. And he found a role for himself among the peace activists: sideshow. Wherever the group landed, Vermin Supreme ran for office. "My character came from the far Right, with policies reminiscent of Jonathan Swift's 'A Modest Proposal,'" he says. He had turned himself into a walking essay.

And he watched as too many peace marches devolved into violence—either because people in the crowd did something stupid, like throwing trash at the police, or because the riot cops lost their cool. So he decided that he would show up at some demonstrations as a clown rather than a candidate, helping to keep the peace. At such events, "I'm a little bit emcee, a little bit town crier, a little bit 10-watt radio journalist, and a little bit Tokyo Rose"—whatever it takes to keep everyone in a good mood. Especially the cops. When a line of police approaches in full riot gear, he whips out his bullhorn and broadcasts reassuring messages to them: "There is no problem here. You are in no danger." Sometimes, he uses his megaphone to lead the police through meditations inspired by New Age relaxation tapes: "I'm going to ask you to seize on your happiest childhood memory. You can feel your breath inside your gas mask."

Or he'll resort to pratfalls. "In D.C. recently, I had a foam tube that looked like a nightstick and I started whacking myself on the head in front of the police, who had their own nightsticks. Whack. Ouch. Whack. Ouch. A lot of them were laughing at me."

Though he has aligned himself with the anti-war movement, Vermin Supreme's true allegiance is to The Prank. And it's this—the anarchism of comedy—that leads him to exploits that seems awfully strange for a man of peace. He recites his attacks on politicians as if he's reading from a resume: "Sometime in the '80s, I bit Jesse Jackson's hand. Also, Jerry Falwell: I jammed a big wad of phlegm onto my palm, and then I shook his hand. I chased Paul Tsongas down the sidewalk and we swung an enema bag in his face."

It's this side of Vermin Supreme that makes me uncomfortable—he seems to have forgotten that even though politicians market themselves as products, even though their hair seems to be made of extruded plastic, they're still human.

Garnick agrees. "I'd call myself a fan of Vermin's, but there are things I wish he wouldn't do. He'll say all these clever things and then he'll go and bite somebody."

VERMIN IS OUR FUTURE

In October, while I was clicking through cable stations, I stumbled across a show called "Who Wants To Be Governor of California?" It asked a collection of real-life fringe candidates to spin a glittery wheel and then talk about whatever issue came up. In the past few years, our country has turned a corner. The political sideshow has moved to center stage. Characters like Jesse "The Body" Ventura and Arnold Schwarzenegger might be called fringe candidates, except that they're winning elections.

"Political disaffection is very high—data shows Americans are more disaffected from government than any time since the polls started recording such numbers," according to Ron Hayduk, an assistant professor of political science at Manhattan Community College. Such alienation, he says, will help to fuel protest candidates and auto-de-fes. Politics is going to get weird. Very weird. It seems only a matter of time before some cable station creates an "I want to be president" reality TV show or a guy in a hot dog suit becomes governor of New York.

What do we want?

Peace.

What will we probably get instead?

A pony.

UPDATE:
Vermin is still running for Mayor of the United States.

The Chemist in the Desert

Two years ago, Dr. Gordon Sato was planting trees in the sandy muck along the coast of Eritrea when the reporters began calling. They wanted to know what Sato thought about the fall of Martha Stewart. Back in the 1980s, Sato had co-invented a drug that helps to block the spread of colon cancer. Now, Martha Stewart was accused of dumping her shares of the company that owned the rights to Sato's drug.

When Sato heard the news, he barely looked up from his digging. "I had no interest in it," he says, of the Martha Stewart scandal. Gordon Sato has one all-consuming interest and that's finding new ways to grow food in the desert. He believes he's on the verge of doing just that. By his count, he and his team of workers have planted over a million mangrove trees in the sand along the coast of the Red Sea, in a drought-wracked, tin-shacked wasteland pocked with overturned Russian tanks, where up until recently not much of anything green could survive. In a pilot program, Sato has shown that the leaves and seeds of the mangrove trees can feed goats, and, thus, provide local villagers with life-saving meat and milk. He says his work will not be finished until he has transformed the coastline of Eritrea into a mangrove park, hundreds of miles of trees. And then maybe he'll move on to other countries. This is an especially ambitious plan given that Sato is now 76 years old.

I am curious about this man who presumes to transform the coastline of an entire nation. What keeps him flying around the world, pushing his solution to world hunger, when so many other men his age are content to hit a few golf balls? And so I arrange to meet him— in Wayland, Massachusetts, of all places. This is where Sato lives in between trips to Hargigo, the Eritrean village that has become his base of operations, and to spots around the world where he meets with people who might possibly donate a few hundred thousand dollars to his project. In a few days, for instance, Sato will fly to Dubai.

Did I mention that he's 76? Sato takes a long time to come to his door. He's deeply tan and wears rumpled work pants, so when I catch my first glance of him through the glass, he strikes me as vigorous. But once I'm inside, I notice the frailty of those sun-beaten arms, the way his clothes hang loose. When he speaks, he pauses, as if marshaling himself for the effort of breath and words. It's nearly impossible to imagine him bumping along in a Land Rover toward a village with no running water. And yet that's where he's most happy.

In his stocking feet, he leads me through a somnolent living room that is bathed in sea-green light from the trees outside. Plump chairs gather around a TV tray, which is set up for Sato's next smoke, all of it laid out just so: the pipe, the matches, the can of Prince Albert. Once we're in the office, the phone rings. He takes the call. He has to. He's a full-time fundraiser now. "This is the job I hate," he says. "But I have to think about a long-term plan, how to have the project go on after I'm dead." Most of the funding for the mangrove project has come out of Sato's own bank account. Still, he has proved himself to be a whiz at raising money. He started his project in 1986 with half a million dollars donated by a Japanese businessman; in 2002, he won the $100,000 Rolex Award for Enterprise. Sato does not bother with the time-honored tools of fundraisers—tact and flattery. "He tends to blurt out what he thinks, often in salty language. But he's feisty and he's smart," according to Bruce Ames, a professor of molecular and cell biology at U.C. Berkeley who has known Sato since they were grad students at Cal Tech. "I think he's a real hero," says Ames. And perhaps most important, Sato has already proved himself right once before, as co-inventor of a cancer drug.

Now he's got another big idea, a plan to turn badlands into wetlands. The secret turns out to be three holes. You take a fist-sized bag of fertilizer and punch a nail through the plastic surface, then bury the bag near the roots of a mangrove seedling that you have just planted in barren soil. Over the next three years, the fertilizer will leak out slowly, supplying the tree with an IV drip of nutrients. If you punch two holes, the tree ends up stunted from lack of fertilizer; with four holes, the tree suffers from too much. Sato discovered this method back in the 1990s, working with Eritrean agriculturalists; he also discovered that goats could live on the mangrove cuttings, though farmers will have to add a cheap supplement in order to fill out the limited diet. "None of this has ever been done before," Sato says.

According to Richard Wright, professor emeritus of environmental science at Gordon College in Wenham, Sato's project could help to showcase the value of mangrove swamps, which are being felled in alarming numbers around the world. "Mangroves provide microhabitats for many species, and that contributes to biodiversity and enhances the stability of the coast," Wright says, pointing out that the trees both hold back the waves and act as nurseries for fish. Sato's project, he adds, "might help people in other regions who have mangroves and have never thought of using them to feed animals."

About seven years ago, Sato began training a team of women to plant row after row of the trees. They've already covered about three miles of the coast, with some of the mangroves now towering over the heads of the workers.

But of course, it's one thing to figure out how to fertilize trees, and quite another to keep any kind of program going in Eritrea, a country studded with landmines, traumatized by ethnic hatreds, and in danger of going to war, once again, with Ethiopia. And then there are the camels. "They're our enemy," Sato says, showing me a photo of a mangrove that has been picked clean. The nomadic people who roam Eritrea let their camels feast on the trees, requiring Sato's team to build fences. "We may be changing the culture of the country because we're fencing off mangroves. The nomads were previously free to go anywhere," Sato acknowledges.

And last year, Sato found out that he had another group of foes, potentially far more damaging than even the camels: coral-reef people. *The New Scientist* ran an article quoting Mark Spalding and other unnamed marine biologists who worried that the fertilizer runoff from Sato's trees—all those chemicals leaching out of all those little plastic bags—could damage the coral reefs off the coast of Eritrea.

"Environmentalists are sanctimonious hypocrites," Sato says, when I bring up the accusations. He pulls at his hair, which is salted black, unkempt and thick, so that it stands up in a ruff, the way a cat's will. Now that he's angry, his exhaustion drops off him. He's leaning forward in his chair, spewing out the reasons why Spalding cannot be trusted, reasons so ad hominem as to be utterly unprintable. Sato asserts that the accusations don't hold up because the runoff from the bags of fertilizer is negligible. "We're not harming the coral reef."

And now I understand what keeps Sato going, keeps him flying around the globe and sends his mangroves marching along the coast. Yes, he's concerned about the Eritrean people. Yes, he's eager to find a direct and simple way to feed the hungry. But what really seems to motivate him is outrage. And in Sato's case, that outrage is more than justified. You only have to see the sign that he's erected in Eritrea to understand why. In English and Japanese, it proclaims the name that he has given to his mangrove project: Manzanar. Sato has taken that awful name, that word with the razor-wire of a Z at its center, and made it his own.

In 1942, the United States opened up its first interment camp for Americans of Japanese descent in Owens Valley, California. The camp, called Manzanar, sat in the middle of a desert. Photographs show tractors plowing through soil so dry that great puffs of powder hang in the air. The people forced to live there suffered from spoiled food, broken refrigerators, bad sanitation. In her book *Farewell to Manzanar,* Jeanne Wakatsuki Houston remembers that "I was sick continually, with stomach cramps and diarrhea. At first it was from the shots they gave us for typhoid, in very heavy doses and in assembly-line fashion: swab, swab, jab, swab, Move along now, swab, jab, swab....Later it was the food that made us sick."

In 1942, Gordon Sato, then a teenager, and his family were deemed "enemy aliens" and sent to live behind the barbed wire at Manzanar. There, he cultivated a small vegetable garden in the dry, dry dirt. That's about all he wants to say about Manzanar. When I ask him to tell more, he shakes his head, and leans back in his chair, into silence.

The boy who dug potatoes in the dust of the California desert would go on to a meteoric career in cell biology. From the 1970s through the early 1980s, Sato worked as a professor at the University of California–San Diego; in 1982, he was named director of a program at the W. Alton Jones Cell Science Center in Lake Placid, New York; he published over 150 articles and was inducted into the National Academy of Sciences. It was in the early 1980s, at U.C–San Diego, that Sato made a key discovery about cancer. With his collaborator, Dr. John Mendelsohn, he found a way to deprive cancer cells of a protein that they use in order to grow. It was this insight that led to the development of the drug Erbitux.

And then in the late 1980s, he retired from cell biology, and devoted himself to answering the question that he says has haunted him ever since Manzanar: How do you make food grow where there's no water? Or perhaps the real question is this: How do you give people who are stuck in some of the worst places of the world a way to support themselves? How do you help a people who have been stripped of their dignity get it back again? Sato believes that enabling Eritreans to grow and harvest their own food, making them independent from handouts, is part of the answer.

In the 1980s, Sato began scouting around for a beleaguered population who lived near a coast and could benefit from an unorthodox agricultural program. He approached Chileans and Chinese groups, but didn't click with them: "They just saw me as money."

Then he called an Eritrean professor in Washington, who expressed interest in his ideas—the professor told Sato to meet him in a room at a nearby university. Eritrea was then at war, and its leaders were deeply suspicious of spies. When Sato arrived at the appointed meeting room, he found it empty. "A phone rang and I picked it up and they told me where to go." He ended up gaining the trust of the Eritrean People's Liberation Front—one of the guerrilla groups then battling the Ethiopians.

From there, Sato flew to Sudan, where he waited in a hotel with his fish in the bathtub—at the time, Sato believed that fish farming was the best way to help the poor feed themselves, but later he discarded the idea as too labor-intensive. A group of Eritrean fighters transported Sato and his fish across the border; a driver had to listen to a shortwave radio every hour or so in order to dodge enemy troops. Two days after Sato left the hotel in Sudan, it blew up.

"But why Eritrea?" I ask yet again. "Why risk your life, when there were so many other places you could have gone?"

Sato shrugs. "I liked the Eritrean people. And I didn't feel it was dangerous at the time."

Whatever the reasons, Sato has found a place that resonates with his own memories, a land of drought and displaced people, of blank sand that hides a terrible past. The village, Hargigo, that is now the center of the Manzanar Project, was once known as a stronghold of resistance against the Ethiopians, the hometown of freedom fighters. In the 1970s, Ethiopian troops tore through

Hargigo and killed or chased out everyone. The place turned into a ghost town. Slowly the survivors and others crept back to resettle the place. Now about 3,000 people subsist here, around a murky watering hole that sometimes turns dry. In this living village, Sato plans to create a massive memorial to the lost town of Hargigo. He wants to plant about 2,000 mangrove trees, each bearing a metal plaque etched with the name of someone who was killed by the Ethiopians. But he has run into difficulty: the village was so ravaged that in many cases even the names of the dead have been lost. This is one of the places on earth so brutal that people do not just get massacred, but also erased.

In the middle of this wasteland, Sato is attempting to build an anti-Manzanar, a place that is exactly the opposite of an internment camp.

"Manzanar might not have been quite as bad as Auschwitz, but it was bad," he says. "I am trying to make Manzanar into something good."

UPDATE:
After this article appeared in *The Boston Globe* Magazine, Sato won a $500,000 Blue Planet Prize. His workers have now planted over one million trees.

One Room, Three-Thousand Brains

When I peer down into one of the buckets in a sink, I see my first human brain. It's actually half a brain, trailing some stem, and the noodle-like folds are pearly in color rather than the gray you'd expect. "We don't want the smell to be too strong," says George Tejada, explaining the need to soak the half-brain in water. It has been preserved in formaldehyde, which gives off a powerful stench.

We're standing in the dissection room of the Harvard Brain Tissue Resource Center, a.k.a. the Brain Bank, which is housed on the McLean Hospital campus. A few minutes ago, Tejada walked toward me, peeling off a latex glove so we could shake hands. "Don't worry, my hands are clean," he assured me. I did not doubt him. Tejada, the assistant director of tissue processing at the Brain Bank, could pass for the headmaster of a prep school, with his crisp button-down shirt and pressed khakis. The dissection room itself, where human brains arrive and get sliced up at a rate of about one per day, also appears to be disappointingly spick-and-span. Aside from the buckets in the sink—and the map of the brain regions taped up above the counter—it would be hard to guess what goes on here.

The institution collects brains from donors and distributes tissue to researchers around the world. The Brain Bank stocks "normal" brains as well as brains donated by people who had schizophrenia, bipolar syndrome, Huntington's disease and Parkinson's disease. It is thanks to the Brain Bank—along with other such repositories—that neuroscientists are beginning to zero in on the genes involved in such mental disabilities. That, in turn, could lead to the development of life-saving drugs.

Thirty years ago, there were no brain banks in the United States—at least not officially. Now, more than a hundred such repositories exist. Over the past few decades, psychology has gone through a monumental transition, away from talk therapy and toward gene therapy. Scientists are sorting, storing and examining human brains as never before in history, looking for clues about why we go mad in tiny slices of tissue. It is a profound shift in the way we think about our own thoughts.

Dr. Francine Benes, director of the Brain Bank, explains about just how powerful this kind of analysis has proved to be. "We're on the threshold of finding the markers for schizophrenia," that is, the genes that contribute to a person's susceptibility to the disease, she says.

It all sounds very worthy and hygienic, but I must admit that I had come to the Brain Bank hoping to be grossed out, at least just a little. My ideas about what might go on at a research facility that houses over 3,000 human brains has been inflamed by an episode of the original *Star Trek* in which day-glo-colored brains ordered slaves to fight battles for their amusement. And then there was an old sci-fi movie called *They Saved Hitler's Brain,* in which the Führer's cut-off head, kept alive inside a glass jar, commands what's left of the Nazi empire. Something about iced brains captures the imagination—after all, no one would bother to make a movie called *They Saved Hitler's Liver.* The brain seems to contain the essence of the self. Yet, unlike so many other dear and familiar body parts—our eyes, our arms, our toes—it exists under wraps. Even as I write that sentence, I'm aware of my own brain dwelling in the loft apartment of my skull, doing who-knows-what up there. It is me, and it is also eerily remote.

"This is a problem that's not going to go away soon," according to Dr. Benes, about the complex feelings people have about their brains, and, therefore, about donating that particular organ to science. "It's believed by many that the soul of their loved one resides in the brain, or they see the brain as what gives one a special connectedness in the spiritual." For years, when people donated their bodies to organ banks, brains were not part of the deal. Now, the taboo against collecting brains has begun to relax. The New England Organ Bank and the New England Eye and Tissue Bank, for instance, have begun to include brains in the roster of donations they collect. The brain is on its way to becoming just another body part.

And that's a great boon to the Brain Bank, in terms of recruiting donors. But outreach is only one small part of what goes on here—most of the work happens after the donor dies, at which time the cells in his or her brain immediately begin to decay. Within hours of getting a call from the donor's family, the Brain Bankers must find a pathologist in the appropriate region of the country, arrange for that pathologist to extract the brain and put it in a special container, and fly the brain to McLean by same-day shipping. After that, the brain will be photographed, assigned a number, sliced up, frozen, formaldehyded, entered into a database, and distributed to worthy investigators.

The brain in the bucket, in fact, turns out to be the only one I come across here that looks remotely brain-like. Every other piece of tissue has been so carefully preserved and processed that you'd be hard-pressed to say what it is. Tejada shows me into a room full of freezers, and opens the door to reveal plastic bags, each of which holds a brain hemisphere cut up into 16 sections. Mist rolls out into the room—the freezer is kept at 80 below. The bags themselves appear to contain flash-frozen shrimp.

When a brain comes in here, usually half of it gets frozen and the other half ends up in formaldehyde—a system designed to give researchers as many options as possible. In the "Tupperware room," slices of brain tissue marinate in chemicals; each half-brain is stored in the kind of plastic container you might use to microwave left-over pasta. From the looks of the original labels, which still cling to some of the containers, the Brain Bank opted for an off-brand rather than genuine Tupperware.

Tejada leads me back to his office and shows me a photo of one the brains—such mug shots are made available to researchers, along with other information, in order to help them select which tissue sample they would like to order. The brain in the photo, freshly cut out of the donor's head, gleams with blood. Unlike the brains in sci-fi movies, this one does not look up to the task of issuing commands to Nazi followers. It's just a piece of meat. I'm reminded of what Williams James said: "The brain itself is an excessively vascular organ, a sponge full of blood." This photo is a powerful argument for using a biological model to understand what goes on in the mind.

"I was part of the shift" toward seeing mental illness as an organic problem, according to Dr. Benes, who is trim and wears a starched lab coat. We're sitting in her office, and she's holding research on her lap—a sheaf of papers covered with numbers that represent the gene profiles of schizophrenic, bipolar, and control-group brains.

In 1973, Dr. Benes attended a neuroscience meeting at a ski resort in Colorado. Then a cell biologist, she had no particular expertise in mental illness. Nonetheless, she attended a talk on schizophrenia, delivered by Dr. Janice Stevens. "I was standing in the back of the room and it blew me away," according to Benes. Stevens proposed that schizophrenia was due to a malfunction in the temporal lobes—a radical theory back when psychiatrists were still blaming schizophrenia on mothers, who were presumed to have driven their children mad with Joan Crawford-like behavior. Stevens' paper helped to erase that stigma. Mental illness became a matter of bad wiring rather than bad mothers. "Schizophrenia could now be visualized in terms of the circuitries in the brain," according to Benes.

The revelation changed the direction of Benes' life. She went back to medical school, and, in 1982, established the Laboratory for Structural Neuroscience at McLean. Early on, "there was me and maybe two other people who were doing postmortem schizophrenia research," she says. By the late 1980s, however, researchers were routinely examining postmortem brain tissue in order to search for the causes of Alzheimer's and Huntington's disease. Using the same methods to investigate mental illness no longer seemed so strange. These days, Benes says, "the stigma is gone and you have the excitement of getting inside the cells," chasing after the genes that lead to mental dysfunction.

Most exciting, brain scientists have begun to borrow ideas from the colleagues who study problems that affect other parts of the body, for instance, the lungs. "At this juncture we're starting to move into alignment with cancer research and hematology research," she says. "That is going to prove to be the most historic step in this field." Treating the brain as just another part of the body—vulnerable to high cholesterol, second-hand smoke, and bad genes—may lead to breakthroughs that were not possible back when the brain seemed to be something special and separate.

***

"I am a brain banker, asking for a deposit from you," Dr. Jill Taylor sings to me over the phone from her home in Bloomington, Ind.

Out there in the Midwest, she's sitting with a guitar in her lap and a phone receiver lying on the floor before her, strumming like a cowboy. "Find the key to unlock this thing we call insanity," she warbles. "Just dial 1-800-BRAIN-BANK for information please." And then she ends with a whoop worthy of Hank Williams.

Dr. Taylor, the Brain Bank's Spokesperson for Psychiatric Disorders, tours the country, exhorting those who suffer from mental illness to donate brains to science. In just about every talk she gives, "there's this wonderful moment when the audience realizes, 'Oh my gosh, she wants my brain,'" Taylor says. "The tension in the room gets really thick. Everyone's looking down like we're all in the first grade—'Don't call on me, don't call on me.' So I'll pull out my guitar and sing the Brain Bank jingle. It lightens everything up. It's been a wonderful marketing tool."

Taylor hooked up with the Brain Bank in 1993, a postdoc working in Dr. Benes' lab, researching the organic causes of schizophrenia. She was determined to prove they existed. Her brother suffered from schizophrenia.

"The thinking in the professional community for decades was that it was a character flaw caused in part by the schizophrenogenic mother—they blamed the family," she says. Taylor fought against that stigma on two fronts: as a Harvard researcher and as a board member of the National Association of the Mentally Ill (NAMI), a group that pioneered the effort to understand mental illness as a wiring problem rather than a failure of the will. "It's only been ten years really that it has been put out that severe mental illness, schizophrenia in particular, is a biologically based brain disorder," Taylor says. In the mid-1990s, Taylor worked in the lab from Monday through Friday, and then would often hop on a plane over the weekend to deliver lectures around the country, spreading the word about brain donation. "When I first started we were receiving fewer than five psychiatric brains a year," she says. Now that figure has jumped to 30.

Taylor admits that back then she was something of an overachiever. "I had been driven towards excellence by very powerful anger. And that anger was related to growing up with a sibling who was not normal. You're constantly on guard. You trust him and then he hurts you emotionally," she said.

On December 10th, 1996, Taylor—then 37 years old—woke up to find she had a brain disorder of her own. "Inside four hours, I watched my mind deteriorate in its ability to process incoming information," she says. A golf-ball-sized hemorrhage had formed in the left hemisphere of her brain, between the two centers that process language. Because the hemorrhage had also knocked out the parts of the brain that create fear, she felt only curiosity as her mental functions shut down. "I learned as much about my brain in those four hours as I had in my whole academic career," she says. "Eventually my right arm went completely paralyzed and that's when my brain said, 'Oh my gosh, I'm having a stroke. I'm having a stroke? I'm a very busy woman. Well, I can't stop this from happening, so I'll do this for a week, learn what I can learn from it, and then I'll get back to work.'"

In fact, recovery took years. "I was an infant in a woman's body," she says, unable to wiggle her toes or roll over in bed. "I couldn't understand when I heard other people making language at me. I lost my ability to read and write." It was two years before she could cook and talk on the phone at the same time, and many more before she could hop from rock to rock without planning exactly where to put her feet.

Last summer, Year Seven of her recovery, Taylor strapped herself into water skis for the first time since her stroke. As a girl, she'd been a slalomer, so comfortable on the water that her skis felt like part of her feet. Now, as the boat pulled away, she found her balance, and began cutting through the glassy water. "All of a sudden there was this moment when my body took its position, and every cell remembered that it was powerful. Off I went. It was pure bliss. I had recovered who I was before that hemorrhage." No longer a weekend warrior, she enjoys a laid-back life in Indiana. She continues to spread the word about brain donations, and she also makes brains—out of stained glass. Taylor's creations are suitable for hanging against a big sunny window, where you can watch the world through the yellow and blue and green pieces of glass that represent the nuclei and gyri and language centers.

***

Right now, in one window of my computer, I'm donating my brain to science, typing information into an online form. It's no more difficult than ordering a book from Amazon.com. You give your name, address, next of kin, and use a pull-down menu to let the Brain Bank know what kind you've got. For a moment, I linger over that list, deciding how to classify myself: normal control, schizophrenia, bipolar disorder, Parkinson's disease, Huntington's disease. I click on "normal control" knowing that designation could change. That blood-soaked sponge in my head, so soft, so frangible, could be destroyed by any of a number of diseases. I have seen the villains up close on a computer screen at McLean. There, a helpful man in a lab coat showed me slides of brain tissue that had been eaten away by abnormalities, from the tangles of Alzheimer's to the bloom of melanoma cells.

We hope, of course, that all of this study of brain tissue will lead to the discovery of new pills and treatments. But along the way, perhaps it will offer another benefit too. The more we know about neurological disorders, the more likely we are to feel compassion for the people who suffer from those disorders. I'm remembering something that Dr. Taylor said about the moment when her brother was diagnosed with schizophrenia: "It was a relief. Finally, I could separate him from the disease. I could forgive him. I could love him again as my big brother."

The Mystic Mechanic

At the Aladdin's lamp, which perches high up on a pole, you turn off the main road and bump along through a back-alley wasteland of Cambridge, Massachusetts. Here, steel statues rise out of the asphalt, wearing coats of red rust and decorated with mingled religious symbols. For instance, a line of menorah candles spells out the word "Allah." The statues seem to come from some parallel Planet Earth where people have stopped warring over God and the only problem left is to come up with a logo for the new, peaceful, one-world religion. Next to one particularly tall steel slab, a cinderblock building squats.

When I walk inside, Mahmood Rezaei-Kamalabad—Islamic mystic, artist, and auto mechanic—is hovering over a samovar. A compact man with a gray-shot beard, he bustles with frantic energy, as if he's the host of a party that's continually just about to begin. "Tea!" he proclaims. "You must have tea!" And he offers a steaming cup. Then Rezaei-Kamalabad, 52, finds himself under a Buick that hangs on a lift, explaining how brain energy affects the fuel injector and telling me about his inventions-in-progress that could help bring the major world religions into harmony. He plans, for instance, to glue a Torah, a Bible, and a Koran into one big tome but has been stymied because the religious texts come in different sizes.

"All my thinking is 'unite, unite, unite,'" he says, gesturing emphatically with grease-blackened hands, a loose Band-Aid flapping off of one finger. He's tortured by contradictions and schisms, so tortured that he spent 17 years and—he estimates—$90,000 to build his masterwork, the Sense of Unity Machine, which he stores in an unheated room behind the auto shop. When he flips on the light to show me, my first thought is "electric chair." A gurney hangs suspended from a steel frame that soars maybe 12 feet in the air. Old seat belts dangle, ready to strap you to the stretcher, and there's a stained pillow to cradle your head. Rezaei-Kamalabad explains how the machine simultaneously spins you around and flips you head over heels, giving you the advantage of two prayer motions in one: You're bowing and also whirling like a dervish. This combo, he says, brings about inner harmony and also cures diabetes. He turns a switch, and the gurney flings itself about in midair, seat belts flailing, while the motor screeches.

In the interests of journalism, I know I should volunteer to strap myself into that thing and see whether or not I come out realigned. But I can't work up the nerve. So instead I scream at Rezaei-Kamalabad over the thunder of the machine: "So, you probably use this machine every day."

He flips the switch off. "No, dear," he says, in the sudden silence. "I used to. But I don't have time now."

I'm startled by this answer: Why would anyone spend more than a decade perfecting a machine and then not use it? But before I can press him on this, the door to his shop tinkles—someone coming for tea. And maybe that's the answer: This garage is also a Sense of Unity Machine, a crossroads for artists, cabdrivers, friends from the public housing development nearby, curious kids, techies, skate-punks, and spiritual seekers. In the hours I spend hanging out at the shop, visitors include a teacher who leaves a silk scarf for Rezaei-Kamalabad's wife and an Indian-born customer who expounds on the history of Christianity in Madras. I find a book on his desk—*DNA: The Secret of Life*—signed by one of the authors, Andrew Berry, a regular at Aladdin Auto Service Center. "I drink tea here," says Ethiopian-born Yohannes Joseph, a neighbor who came that day just to hang out. "I have tea in my house, but this tastes better. This place is like our home."

In the 1970s, Rezaei-Kamalabad worked for the government under the shah of Iran, overseeing the repair of palace buildings. In 1978, he left for Boston, inspired by a message from Allah—who directed him to pursue an American education. He got out just in time. In 1979, the Ayatollah Khomeini's revolution ripped Iran apart. "If I was there during the revolution, I could be dead," Rezaei-Kamalabad says simply.

He drove a cab and studied at Massachusetts Art, where he fell in love with a fellow student, Marianne Murphy. They married in 1984. He wanted to return to Tehran and live there. She didn't.

One night, he says, he prayed for guidance—should he stay or go?—and God sent him a dream: He found himself gliding down the Charles River in Noah's ark, fog curling off the water. In the gloom, he could just make out the lights of the Hyatt Regency hotel. He knew the place well; every day he picked up customers there in his cab. The ark stopped in front of the Hyatt. Its door opened. He was sent out onto the shore, alone. When he woke up, he knew he had to stay in the United States. "I start to cry," he says. "I did not have any roots here. Everything was in Iran."

A few years later, he began building his Sense of Unity Machine, which would rock him in the air, would cradle him in its curling prayer, bringing harmony to a man living in exile. When I think of him hunched over those plans, I'm reminded of Temple Grandin, a high-functioning autistic and Colorado State University professor of animal science who invented a machine to hug her, because she could not stand the touch of another person. Did Rezaei-Kamalabad hope for a machine that could soothe his particular pain?

Maybe gyrating through the air—as he used to do, back when he had time to ride in his machine—worked some kind of magic on him. He exudes a rare sweetness.

"Do you have an e-mail address," I ask him, "so I can send you your quotes? I don't want to misquote you."

"Oh, no, my dear!" he replies. "I would never interfere with your art. You can have me say whatever you want. Have me say swear words!" He laughs uproariously at his own joke and then hurries to his samovar, to pour in more spices.

The Ballad of Conor Oberst

Conor Oberst takes the stage in Eugene, Oregon, looking like a cartoon of an indie rocker. His hair flops around his head in a black bubble, and above the quickly drawn lines of his mouth and nose his soulful eyes stare out. Though he's from Nebraska, he brings to mind some Japanese pop icon—Hello Kitty perhaps—the way he manages to seem both adorable and haunted at the same time.

Oberst, once the child prodigy of the indie scene, has always had to battle his own cuteness. He wrote his first songs at age 12, began soloing in clubs at 13. Now, at 22, he has played in so many bands and been involved in so many side projects that you need a flow chart to keep track of them. He has confessed crushes into microphones; he has shaken and shivered; he has built a following as the bard of teenage angst. Until this year. Suddenly, Oberst has swerved in an entirely different direction. On a recent CD he made with Desaparecidos, one of the bands he leads, he lambasted American consumerism with a level of sophistication you rarely hear in three-minute songs. And in Bright Eyes—the band that's clearly closest to Oberst's heart—his writing shows a sudden leap toward maturity.

"Onto the stage I was pushed with my sorrow well rehearsed," he sings on the band's latest CD. "I hear the ice start to melt and watch the rooftops weep for the sunlight./And I know what must change." With these two literate albums out this year and a knock-out performance at the recent C.M.J. festival in New York, he has critics buzzing that he might be the next Bob Dylan. But for the kids who make up the audience tonight, Dylan is grandpa. The club smells like a high-school gym, of teenagers in heat. The kids are hip in the studied way of those who take advanced-placement English classes. Oberst thanks everyone for showing up. He shuffles his feet, as if overcome by a fit of modesty. He chugs from a bottle of wine.

He prefers not to stand stark alone on the stage, so Bright Eyes contains an ever-changing bunch of friends. On this tour, Oberst has surrounded himself with more than a dozen bandmates, a glockenspiel, hammer dulcimer, bassoon, violin, cello, drums, guitars, banjo. It's so crowded up there that the musicians have to jump over instruments as they navigate the stage. From song to song, they transform themselves from countrified hootenanny to folk strummers to guitar heroes—it's hard not to be awed by their range. Oberst melts into the background, popping out now and then for solos.

Truth be told, he does not seem the least bit like Dylan. He jangles his guitar strings and sings: "Love is an excuse to get hurt

and to hurt./'Do you like to hurt?'/'I do. I do.'/'Then hurt me.'"
He fixes us with those Pokémon eyes. The kids—some wearing
glasses smeared with reflections from the pink lights of the stage—
mouth the words to his songs.

And then, about halfway through the set, the mood shifts. The
band wanders away. Now it's just Oberst up there, and he leans into
the mike to sing "Don't Know When but a Day Is Gonna Come."
In the recorded version, Oberst mixed his voice so low you could
hardly hear it, as if embarrassed by his lyrics. Now he hurls them
out. "They say they don't know when, but a day is gonna come,
when there won't be a moon and there won't be a sun. It will just
go black." Each line lands like a piece of prophecy. As Oberst rips
into a verse that wasn't on the CD, the audience goes silent. They
hold their breath.

He's angry. Oberst—sweet, skinny Oberst—is so angry that he
hisses the words: "It's hard to ignore all the news reports. They say
we must defend ourselves. Fight on foreign soil. Against the infidels.
With the oil wells. God save gas prices."

A boy holding a bike helmet curls into himself, punched by the
words. The girls doing the bump freeze, their hips still cocked.
Then the crowd shouts with delayed delight, almost as if fireworks

have popped up near the rafters. Almost as if Oberst, by boxing the war into a rhyme, might be able to stop it.

The next day, Oberst totters out of the Eugene Hilton complaining that he's been poisoned. Too much wine last night. He's detoxing with a supersize coffee as he drags along the sidewalk beside me. Up close, Oberst loses his cartoon-character glossiness. He is alarmingly pale. His hair falls into his face in loops, and he hides behind it. He begins a sentence, lets it trail off. He interrupts himself, stumbling over a thought. He's skinny in a way that is only possible for a 22-year-old rocker on the road, with black-jeaned toothpick legs and a chest shaped like a fifth of gin. During the three days I trail his band, the only items of food I see pass his lips are a few pistachios. Admittedly, I am not with him some of the time. Still, his seeming ability to live on nothing but a tiny green nut or two contributes to his ethereal presence. He exudes the otherworldly aura of a person who—in some back part of his brain, some Omaha of the mind—is always writing a song.

I'm escorting him to his next gig, Portland, in a rental Chevy. The young Bob Dylan would have made a miserable companion on a car trip—bitingly sarcastic (remember that scene in *Don't Look Back* when he tortures Donovan?) and perpetually on the make. Conor Oberst, on the other hand, helps out by reading the map and asks permission to turn down the air-conditioning. Seemingly unaware of the stereotype concerning men and directions, he cheerfully rolls down his window and leans out to query passers-by, "How do we get back on Route 5?"

When he's not map reading, Oberst hunches over a newspaper. The news, as usual, is terrifying. North Korea is boasting that it will soon possess nuclear weapons. "I was sort of getting used to Iraq, and now this," he says.

I ask him when he decided to add the verse about Iraq that he sang the previous night. He tells me that he originally wrote the song that way, but then cut the most tendentious words out, worrying that it might sound "preachy." During this tour, which coincided with the congressional debate on Iraq, he decided to put the verse back in. On most nights, his show peaks when he surprises his fans with the new-old lyrics.

"I thought I would never sing political songs, but now it seems like the only thing worth singing about," he says. "It's difficult to think about anything besides the war." He tells me—only half-joking—that he's come up with a plan in case the draft is reinstated: he'll marry a Canadian. Of course, in the unlikely event of a draft, middle-class poets like Oberst would be the last to fight. He slides his eyes toward me shyly. "I always embrace the worst-case scenario," he admits. For some reason this strikes both of us as a punch line, and we crack up.

What must it have been like to be 21 when the towers fell? Oberst and his gang used to measure themselves by the record deals they made or turned down. They worried about how to stay raw, about—really—too much prosperity. Now, according to one of Oberst's bandmates, they agonize about a friend in the military who may be shipped to the Middle East. They have vague plans to take their protest beyond lyrics, but they're not sure what that would mean.

Oberst swam through high school during the plush years of the 1990s in circumstances that could be described as swaddled: he attended Catholic prep school in Omaha and hung out on what he described in one song as "that perfect peaceful street." His dad, who works for Mutual of Omaha, played classic-rock songs in a wedding band; his brother Matt jammed with the neighbors; photos on a CD of songs from 1995 to 1997 show his parents kissing, kids playing with a reel-to-reel tape machine lit up in the background.

As a seventh grader alone in his room, Oberst began to pour out songs into the four-track, recording tortured moods. David Verdirame, a former classmate of Oberst's whom I met backstage at a show, remembers that kid with the sophisticated musical taste. "I was wearing a Nine Inch Nails T-shirt, and he made fun of me. He made me a mix tape of all these local bands I'd never heard of." Verdirame—now a well-scrubbed fellow who could pass as a frat brother—locks me in an intense gaze as he tries to describe what the discovery of edgy music, especially Oberst's, meant to him.

"I had a lot of old tapes of Conor's music," he says. "They were so disturbing I had to throw them out." His blue eyes laser into mine. "I threw them in the ocean."

Mostly, though, Oberst hung out with older musicians, who treated him as a peer. One "was playing a show one night," Oberst remembers. "I went down to see him, and he was like: 'I'm not going to play my last song. Conor's going to play.'" The 13-year-old took the stage. Soon he was performing around town, screaming and flailing as if to exorcise his teenage demons.

Meanwhile, Oberst formed a band called Commander Venus. They recorded for a small label that eventually turned into Wind Up Records, which turned into one of the largest indie labels, home of the stadium rockers Creed. Before he had graduated from high school, Oberst learned what it was like to interact with big business. "I don't think I could ever be on a major label," he says, "because their idea of success and mine are vastly different."

When Commander Venus broke up, Oberst and his friends decided to form their own label, which started out almost as a hobby but grew into a real business. "The label splits the money 50-50 with the bands," Oberst says. "There's no paperwork. It's trust." They named their outfit Saddle Creek Records, after the Omaha street where they'd hung out as kids.

The share-and-share-alike ethic carries over into the way Oberst puts together his tours. Bright Eyes travels as a kind of summer camp on wheels. When Oberst tells me how he lost all his clothes in a New York cab a few weeks ago, I point to his faded black jeans, the long-sleeved T-shirt. "So where'd those come from?"

He ducks his head, a bit sheepish. "My bandmates," he says. "We always wear each other's clothes." I think back to when I was that age, how boyfriends and girlfriends would announce their passion by trading jeans and sweaters, parading around in costume as one another. Oberst must be in love with his whole band.

Later that day, in the back of a Portland club, the band's bassoonist, its bassist, a filmmaker and the guy who sells CDs at shows hang out in the kitchenette. Other band members pound up the stairs and find beers, then vanish. Somewhere downstairs a drum rattles experimentally. A banjo twangs and then stops mid-riff. With 15 musicians and a bestiary of instruments, the Bright Eyes sound check starts at 3 and drags on for hours, sometimes right up until an 8 o'clock show. Now, late-afternoon light cuts through a window, illuminating the attic dust that meanders lethargically through the air.

The club has taken on the atmosphere of a sprawling group house in Omaha. Someone suggests that Oberst should cook everyone dinner in the kitchenette—and then the conversation winds its way back to other dinners whipped up in other cities. Oberst himself wanders in, sprawls on a stool, listens, laughs appreciatively. You can catch in him some echo of that 13-year-old boy who hung out at college parties, the perpetual little brother just happy to be included. You would never guess that he has masterminded all of this.

I have been Conor Oberst. And you can be him, too. One of his fans has created an online game that lets you inhabit Oberst's body (albeit a very crude, pixelated version of his body) so that you can lob seven-inch records at Steven Tyler, the members of Creed and Fred Durst until you eventually destroy them. I contacted the inventor of the game, 23-year-old Jason Oda, to ask him what exactly prompted such a labor-intensive tribute to his hero.

"I've never seen anyone in music be so tormented," Oda said. "The song about the coughing, shaking fit on the bathroom floor—I don't know if I'm sounding teenaged, but when you are drunk and passed out on your bathroom floor and screaming out loud and no one can hear you because your apartment's lonely and cold, it's the perfect music."

Because of such raw emotionality, fans and reviewers used to peg Bright Eyes as an emo band—"emo" being a branch of hardcore punk in which it's O.K. for boys to cry. Oberst feels no allegiance to emo; in fact, the word has become a slur lately, especially after *Seventeen* magazine asked "Am I emo?" in a spread featuring a girl model with black-dyed hair and a boy model in nerd glasses and a grandpa sweater.

The question for Oberst is not "Am I emo?" but "Am I pure?" Even now that he's preoccupied by global politics, he continues to be fascinated with his own inner landscape. Because if you produce exactly the songs you want to for your own label, even then, the pressure of fans' expectations can push you toward fake sentiments and slick hooks, songs that are nothing like what you would have written if you'd been strumming in your basement with guys from down the street, back in those days when you had no expectations beyond playing at a coffee shop in downtown Omaha.

Oberst goes through gyrations to stay true to that original spirit. For one thing, he gave the 2002 Bright Eyes CD a long, Dickensian title ("Lifted or the Story Is in the Soil, Keep Your Ear to the Ground") that makes it hard to list on a sale rack. For another, he chose to begin the CD with a song that's almost painful to listen to, a cracked voice backed by desultory guitar-strumming—you must tolerate the broccoli of authenticity before you get to the treats that follow.

Critics have raved, though some wonder if Oberst is spreading himself too thin. Peter Choyce, a longtime college-radio D.J., praises Oberst as an "icon" but wishes he would edit himself. "Every time I go to the station there's more product. Him with the guy from Spoon. Him in Desaparecidos. Him with a new EP. I can't keep up. Better to release one strong, solid 10-track CD a year and make people want more." It is, of course, exactly this kind of logic that Oberst resists.

On tour, he writes songs in his motel room, on the bus, while everyone else is napping off hangovers. The band has incorporated one of the just-finished songs in its set, a daring move.

"If there's a song that stops meaning anything to me, then I'll quit playing it," he says. And if a song does mean something, he'll stick by it, as he has stuck by the friends he has known since he was 14, whom he has turned into his bandmates and collaborators and business partners.

In the winter, when Oberst and his friends decided to release a CD of hard-rock songs, they did so under the moniker Desaparecidos, naming themselves after the disappeared activists in Argentina. "Read Music/Speak Spanish" is an odd and amazing artifact, a rock album that examines the sociopolitics of urban sprawl. Instead of condemning the suburbanites who have chosen to live behind vinyl siding, Oberst's lyrics imagine the mental life of America's disappeared: "If you need money for bills, my lover I could cover you....I'm saving up. We can get that house next to the park. I'll get more hours at my dad's shop, yeah we'll plan for everything. And we'll enroll in that middle class. Get a compact car full of discount tags. If you're feeling trapped or too attached remember we wanted that." Oberst has created a Raymond Carver story in 4/4.

Though the lyrics seemed taboo in that flag-waving period after 9/11, the band released the CD anyway. "It was the worst time to be singing anything unpatriotic," according to Oberst.

Casey Scott—a tall Georgian with shaggy hair and a Southern Comfort drawl—played bass with Desaparecidos on tour and has toured with Bright Eyes. He is also Oberst's housemate back in Nebraska. The fans, he says, used to be "college English majors," but in the past year all that changed. Now Oberst hides his identity when he signs into hotels so as not to be stalked by histrionic 14-year-old girls. "People are always giving him stuffed animals. Because what he writes is so honest, they feel like they know him personally. They tell him the craziest things that they would never tell anyone else."

We're sitting in a cafe in Seattle. Across the street, kids mob the entrance to the club where Bright Eyes will play in two hours, looking from this distance like bees hovering around a hive—they're not lining up so much as waiting for a glimpse, a stray band member who might venture out of the green room. Whenever I glance out the window, there's more of them—girls with pink hair, bespectacled boys, black thrift-store overcoats, Dr. Seuss shoes.

Scott says that until this year, he never bothered to tune into the news much, but he's scared now. He's scared of the world. "Even if I didn't know Conor, by now even I would have taken notice" of the looming war in Iraq, he says.

Outside, a line has formed, stretching down the block. Kids sprawl on the concrete, drape themselves on the side of the club, take up room the way only high schoolers can. Even from here you can feel the suck of their longing, the weight of the secrets that they dare confess only to Conor Oberst. Maybe years from now they'll be known as members of the generation startled out of puberty by 9/11. Or maybe we will know these kids, or their peers, as the ones who fought in the streets of Baghdad. But one thing is clear: if any generation ever needed a new Bob Dylan, this is the one.

How to Make (Almost) Anything

In 2000, Saul Griffith—then a grad student in the MIT Media Lab—traveled to Guyana as a volunteer. Working with a retired optometrist, also a volunteer, he handed out recycled eyeglasses to people who would otherwise have none. The glasses had been collected in donation bins by a do-gooder organization in the United States and refurbished by prisoners. This, unfortunately, is still the state-of-the-art method for delivering eyeglasses to poor people in remote villages.

"It was terribly depressing," Griffith says. He remembers particularly the young road-crew worker who came to him one day. The fellow had waited a long time for glasses; a line of about two hundred people stretched out from the door of the community center, people patiently shifting from foot to foot near gutters clogged with water lilies. The guy had terrible vision, though otherwise he was the picture of health, a handsome charmer, in fact. Griffith searched through boxes for something that would address the man's nearsightedness and astigmatism, and he found just one pair that would work: pink cat-eyes with rhinestones at the tips.

The man laughed and said, "I'll never get laid in those." Then he left without his glasses, walking back out into the smeared sunlight and indistinct streets with his pride intact.

That's when Griffith decided it just didn't make sense to try to match people with recycled glasses. He estimates that each pair he fitted cost about a hundred dollars, if you figure in airfare for volunteers, jeeps, refurbishing, shipping, etc. He thought it could be cheaper to bring small factories to rural villages, and to make eyeglass lenses on the spot. "Usually, you need 2,400 prescriptions to cover a population. It's terribly hard to have that inventory" in a remote village in Africa or India, he says. "I asked myself, 'What about a manufacturing system? Can I allow someone to be their own manufacturing center and service 10,000 people?'"

You might say that he was exactly the right person to explore this question. At the Media Lab, Griffith belonged to a milieu—and indeed, helped create a milieu—of designers who are excited about the advent of cheap manufacturing.

Neil Gershenfeld, a professor at MIT's Center for Bits and Atoms, calls them "personal fabrication" machines—affordable devices that will let us mold anything from plastic to computer chips. Gershenfeld already uses one. By hooking together a series of off-the-shelf fabrication tools, he has created a machine that can itself make machines. "This technology is here today," he says of his rig. It costs about $20,000 for all the equipment, but prices are dropping quickly. With a milling machine, a laser cutter, a vinyl cutter, a PC and some "glue software" to let the machines talk to each other, you've got enough power to etch your own circuit chips and sculpt any flat shape you want out of wood, acrylic, copper, or what-have-you. Gershenfeld encourages MIT students to dream up new uses for such fabrication tools in his popular class, "How To Make (Almost) Anything."

He has also set up one of his machine-making machines at the South End Technology Center in Boston, a community center where anyone can drop in to learn computer skills. There, he says, "eleven-year-old girls are making circuit boards." The kids use a PC and a milling machine to hardwire instructions into logic chips, and then they drop those chips into their own rudimentary gizmos, for instance, a Pac-man shaped joystick.

"I have a student," Gershenfeld says, "who will graduate when his thesis can walk out of the printer."

***

Griffith is frying eggs, gingerly holding the spatula in his injured hand. A wound runs like a dotted line across his palm and fingers. Kite-surfing accident, he says. Griffith invents or alters much of his own sporting equipment, a collection of Mad Max skateboards with fat tires, as well as kite boards made out of plywood planks and pink insulation foam from Home Depot. Gashes come with the territory. On his web site, he has posted extreme close-ups of one of his past wounds, a slice that appears to go deep into his foot.

Eggs done, Griffith crashes into the living room, shoving away papers, bits of wire and dirty mugs to make space on the table so he can eat his breakfast—I've come by his place at nine in the morning, which is painfully early for a guy in graduate school. When he's finished, we take a look at the eyeglass-lens printing machine, which sits in a closet just off the living room. Griffith has just invented this suitcase-sized factory; it makes eyeglass lenses for any prescription. And it could change the lives of one billion people in the world who need glasses but don't have them.

"This is the bottom half of the machine," he says, pointing to a device the size of a VCR. In its center, a piece of silver foil stretches over a circular frame. Griffith works a hydraulic pump, sucking fluid out from under this foil, which dimples downwards to form a lens shape. You can buy the foil at your local auto-parts store: it's the stuff used to tint windshields. The fluid that Griffith pumps in and out is baby oil—and it gives the machine its own particular aroma, a faint smell of beaches and tanning.

Here's how the machine would work in the field: When the client hands you a prescription, you adjust the foil to the appropriate curvature, pour liquid acrylic into the resulting lens mold, wait a few minutes for the acrylic to harden, and—presto!—you have the front slice of the lens, the part that corrects near-sightedness or farsightedness. The back piece of the lens—the part that corrects astigmatisms—gets molded in a similar baby-oil contraption. A gallon of liquid acrylic costs about nine dollars, and a gallon goes a long way, so the resulting lenses would be very cheap.

It's the world's first machine to print eyeglass lenses on the fly, and the feat has caused something of a sensation. This year, Griffith won the MIT-Lemelson Student Prize—one of the most prestigious awards for young inventors. He has since fielded hundreds of e-mails from people who want to buy the machine right now. The trouble is, Griffith only has one machine, this one, and the road from prototype to shrink-wrapped package will be long and expensive.

"People call you up and say, 'Oh we love it,' and waste hours telling you how wonderful your idea is. You say, 'All I need is money to get it from here to product' and that's when the call ends." It would cost Griffith from half a million to five million dollars to get his product to market. When you're designing tools for people who live on $2 a day—it can be especially difficult to raise that kind of money for research and development.

But Griffith has a plan: call it his first-world strategy. "Bifocals, trifocals, and progressives," he says. Baby boomers need glasses that are customized for hobbies that involve both close and far vision. "My father plays golf and the progressive glasses that he uses in his office don't work for a golf swing," Griffith says. "You can imagine people having different progressives for all their different activities," from photography to knitting to gardening. With a few tweaks, Griffith says, his machine would be able to spit out customized lenses for fussy first-worlders, and that might help him fund the machine he really wants to make, the one for people who have no glasses at all.

"Eyeglasses are the ultimate in personal fabrication," he says, "because everyone has different needs." It's the eyeglass lens, that little chip of plastic that comes in thousands of curves, that argues for Griffith's philosophy: We should be able to mold the things in our lives into any shape we want.

***

Griffith is showing me one of his favorite books, *The Boy Mechanic.* When he himself was a boy mechanic in Sydney, his mom (artist) and dad (textiles engineer) presented him with this how-to guide aimed at junior Edisons, originally published in 1913. It was one of the few such books they could find. "Why hasn't another book like this been written in the last hundred years?" Griffith asks. "Maybe because of liability."

Liability indeed. He opens up the cover to show me the illustration on the inside leaf, an engraving of a boy about to fly in a full-sized glider plane, which looks to have been patched together out of canvas and sticks. "They're telling you to build a plane and you're twelve years old," says Griffith. "Look at this. You're supposed to jump off a cliff." He stabs his finger at a smaller illustration below. Now the boy stands at the top of a precipice. A dotted line shows his intended trajectory—over a steam train, across a river, and into some bushes beside a house.

The Boy Mechanic also gives instructions for rigging a Ouija board to produce messages from the spirit world, counterfeiting money, and using a pipe bomb to skyrocket a mannequin high into the air. This was science education a hundred years ago. Girls did not exist at all. And everyone assumed that boys would be jumping off cliffs and blowing things up anyway, so why not teach them to build a steam engine while they're at it? The book prepared young men for a world in which they would have to know how to handle stuff—they would fix their own tractors, build houses, grow food in the backyard. But over the last century, we've turned soft. We've grown awkward with hammers and looms, screwdrivers and sewing needles. We've become strangers in the land of stuff.

Griffith plans to change all this. He hopes to help transform the next generation into tinkerers, boy and girl mechanics who will use a new class of desktop factories to design their own toys and, eventually, the world around them. "Adults are a lost cause. You've got to get the kids. You've got to do a Catholic church on them," he says.

Personal fabrication machines could be part of that transformation. He points out that the first generation of kids who grew up with home computers—Bill Gates, for one—learned to be software wizards and hackers. They understood the potential of digital code as their elders never could. Griffith believes that when kids get hold of milling machines and laser cutters and 3-D printers some similar magic will happen. The next generation, he believes, will be the stuff hackers.

A year ago, Griffith found a new way to advance that cause. One day his friend Joost Bonsen, a graduate researcher at MIT, popped into his office with a big idea. "What we really need to do is create cartoons for kids and show them how to build things," Bonsen remembers suggesting. By the next day, Griffith had roughed out a sketch for a one-page cartoon, which has since evolved into a series called Howtoons. In it, a cast of kids brainstorm better methods for lobbing marshmallows, shooting puffs of air, and covering their bodies in duct tape. While there's some minimal plot in each strip, mostly these rubbery-looking tykes demonstrate how to build toys out of stuff you can find in the Dumpster—soda bottles being the favorite material. "We have an aspiration to put Howtoons on the back of soda bottle labels so that kids will think about how to continue to use the bottle instead of throwing it out— as a jet car, or jet boat, an underwater periscope, or a water-droplet microscope," according to Bonsen. While the cartoons do not advocate blowing things up, they retain something of the cavalier spirit of *The Boy Mechanic,* as well as a muscular quality that strikes me as thoroughly Australian. For instance, one cartoon strip encourages kids to drill through steel, file a blade, and then skateboard across the ice with a sail (wearing a helmet). Griffith and Bonsen lab-test the projects and co-write the text for each cartoon. A professional illustrator does the drawing.

Howtoons is not just a side project for Griffith—he believes the cartoon could be very, very big. Griffith and Bonsen are now negotiating a contract with a major publishing house that would release Howtoons as a kids' book. They've hired a lawyer to hash out the TV and film rights. Currently, they post the cartoon on the Web (www.howtoons.org), and both Griffith and Bonsen want to make sure that any deals in the future won't stop them from disseminating the cartoon for free to kids in developing nations, perhaps with the food and clothes that non-governmental organizations hand out. "Our founding goal is to get these Howtoons to everyone everywhere. We're talking about translations—the top ten languages in the world. The first Howtoon has already been translated into Portuguese," according to Bonsen. Every kid, they believe, should have access to MIT-style experimentation.

It is this kind of fun—or, at least, the promotion and packaging of it—that may end up being his greatest invention.

***

In a back wing of the MIT Museum, the part that's closed to the public, a half-dozen kids carom around between a Pac Man machine, a chair made out of a shopping cart, and floor-to-ceiling shelves filled with trash-picked motor parts, soda bottles, lengths of rubber, PVC tubing, scratched CDs, copper wires, magnets. This is the Howtoons club house. Every few months, Griffith and Bonsen throw a pizza party in this cluttered cavern, invite their friends to bring kids and test out projects that might be worthy of a Howtoons cartoon. Tonight Griffith stands at the front of the room, jumping around and waving his hands until he gets the attention of his three-foot-high colleagues. He asks them what kind of machines they would like to invent. A squirmy fellow of about five years raises his hand. "Spirits," he says. "We could make animal spirits."

It's one of the few requests that Griffith and Bonsen can't accommodate. A few months ago, when a kid suggested a hovercraft, partygoers jumped right in and built a "hoovercraft" out of a Toro 220 leaf blower and a wood plank, with a tarpaulin skirt that trapped a cushion of air underneath. The hoovercraft was powerful enough to float Bonsen and Griffith across the room.

Tonight, I watch as two girls, both about six, dab chocolate frosting onto Cheerios and drop their creations, along with some uncoated Cheerios, into a bowl of milk. A mom shakes the bowl gently to get the cereal circulating, and the girls watch, barely daring to breathe, as the pale Cheerios float around and seem to find one another, sticking together in clumps. The chocolate ones do the same. (The Cheerios' behavior is based on a simple principle: oil and water don't mix.) If you paint each Cheerio with stripes, you can get even more complicated patterns to bloom in your cereal bowl, flowers and snowflakes, because you have "programmed" your Cheerios to touch each other only at certain points of contact.

Griffith came up with the idea. It is, in fact, a kiddie version of his PhD thesis. This spring, he put the finishing touches on a demonstration involving an air-hockey table and Lego-like plastic tokens that float on its surface. The tokens skitter around the table, sometimes sticking to each other and sometimes not, slowly resolving themselves into strings of yellow-yellow-blue. Each plastic token contains a logic chip that tells it when to latch on to a blue or yellow partner, and these instructions, taken together, allow many small things to build themselves into one big thing. It's an insight that has implications in nanotechnology, where subatomic stuff must be pieced together to form bigger stuff. "The way we make everything now is top-down. Machines can only make things that are smaller than themselves. But why not go in the reverse direction and build something bigger than the machine? DNA does it," Griffith says.

He may have proved his point on the air-hockey table, but there's still some technical glitches to be worked out in the cereal bowl. As the girls watch, chocolate begins to flake off the Cheerios. The self-assembling machine is turning to mush. The girls don't seem to care. They've been distracted by a box of Cocoa Puffs, which has been put on the table to encourage further science experiments; they're reaching out to grab handfuls and stuff their mouths.

"No Cocoa Puffs," a dad says. "You're about to have dinner."

"Please?" one of the girls says. "Just one?"

"No."

"Please?"

For a moment, those girls looked into a cereal bowl and saw science. But now all they can see is the sugar. Griffith has big plans for this generation. He can only hope that his plans fit into their agenda.

UPDATE:
Saul Griffith won the MacArthur "genius" grant in 2007.

What We Mean By Freedom:

Three ways of looking at alternative fuel

MY NEIGHBORHOOD, MY TREE FARM

One day I came up with a scheme to power our house on trash. It was in the middle of summer, and I had just sawed up apple-tree branches and stuffed them into bags; now I was dragging the wood out onto the curb. My boyfriend and I live in a Victorian house in the middle of a city, with nothing but a teeny patch of land around it; nonetheless, our yard generates an enormous amount of timber. And now I was throwing it all away. When I looked down the street, I could see yard bags lined up everywhere; our neighbors were dumping potential firewood too. That was when the "aha" moment came: Why not turn my neighborhood into a tree farm?

In a fit of enthusiasm, I convinced my boyfriend that we needed a wood stove. I explained that we could grow and harvest trees on our own postage-stamp yard, and we could trash-pick the rest from the neighbors. He tried to mount arguments, but I was not to be talked out of my idea.

Much drama ensued. A craftsman named Vaclav spent days on his hands and knees in our dining room, using old-world techniques to wrap a slate shield into a corner. Zoning laws had to be consulted; a building inspector needed to be charmed. The stove itself—black, warty-looking, and insanely heavy—had to be fetched from a faraway store. It was one of the cleanest-burning models I could find, and even contained a catalytic unit that would re-burn smoke.

By mid-winter, the stove squatted in its corner, eating like a hog. We fed it wood we'd collected from our yard and our neighbors'; it scarfed the supply in about two days time. And then we had to buy it a huge stack of firewood, delivered to us by a truck that vomited diesel fumes. Things weren't exactly going as planned. The project had ended up being far more expensive than I can admit, even to myself, and our neighborhood made a lousy tree farm. Still, that stove worked like charm. The downstairs of our 19th-century house used to be so frigid in the winter that we had to wear coats and hats while we cooked. Now we lounged in a ski chalet.

This past summer, I became obsessed with perfecting my scavenging techniques, so that we would get through the winter on free wood. Had I been a farmwoman in the 1850s, I would have tramped out into the forest and collected armloads of dead wood. This being the 21st century, I decided that I would hunt on the Internet. In the Craigslist "free" section, I typed "firewood." Bingo. I found a woman in a nearby suburb with the remains of an entire oak in her yard; the tree had been chainsawed into logs, the largest of which probably weighed eight hundred pounds. My boyfriend and I took two carloads of the smallest stuff and barely dented her supply—if we'd owned a truck we could have amassed enough fire-wood to last for years.

Then my boyfriend and I spent all summer hand-sawing that wood, along with branches from our yard, into stove-sized chunks. We developed really nice-looking biceps and shoulder muscles. It was the sawing—not the amount of wood—that ended up limit-ing our supply. Why didn't we buy an ax and chop the wood? Because we have more in common with Paul Krugman than Paul Bunyan, and we'd be likely to get into a political debate while we chopped and end up losing fingers or toes. Still, we managed to produce enough wood chunks to get us through the winter.

But how green is our wood heat, really? Our newfangled stove burns about 90 percent cleaner than would its forebears in the Victorian era. When you look up at our steel chimney, blazing against the sky on a cold day, you can't see much smoke, just a tiny smudge of gray. Even so, that smudge does carry a significant dose of lung-killing particles. Experts argue about how much particulate matter a wood stove generates in comparison to diesel, but it's almost certain that our super-efficient stove is dirtier than an oil burner. The wood stove wins in other ways. It eats trash that comes from our own region. And it is a carbon-neutral energy source, since the CO_2 produced in wood smoke is absorbed by trees, and so on, in a sustainable cycle. Petroleum fuel, on the other, has to be extracted and hauled thousands of miles; it also pumps carbon into the atmosphere that would otherwise stay locked under the ground.

I latched onto wood heat for its geopolitical appeal; I wanted to find a fuel source that had nothing to do with the war in Iraq. But in the end, the wood itself seduced me. I feel in love with it: the cutting of it and the burning of it.

When you pick up a split piece of oak, it's heavier than you expect; it feels in the hand like an encyclopedia from the library of your childhood. Maple cuts easily; you find yourself wanting to stroke its sinuous skin and surprisingly obscene crotches. Apple bleeds when you saw it; on the inside it's orange and pink as a tequila.

The split pieces flame fast and sloppy, while the logs bake slow and mellow. My boyfriend and I part with each hand-sawed, hand-gathered piece a little reluctantly. And, when the stove really gets going, we close the pocket doors that divide the living room from the dining room, shutting ourselves in with the stove, to roast like chestnuts.

(a version of this piece was published in *Dwell* magazine)

THE GREASE QUEST

"This is really the chunky stuff," Jason Carven says, lifting up a plastic container of brown ooze. "Without a doubt, you've got to pump it through a filter." We're standing in his warehouse space, surrounded by silver fuel tanks. It's freezing in here. Over many sweaters, Carven wears the kind of canvas jacket favored by guys at gas stations. With his wan, bearded face and flashing spectacles, the 25-year-old looks like a tubercular poet in drag as a mechanic. He is, in fact, the inventor of the Greasecar Vegetable Oil Conversion System, which will turn an ordinary diesel car into an ecologist's dream, available for $795 over the Internet.

He leads me into another room, where he stores the filtered vegetable oil and selects a two-gallon jug. In the parking lot, snow whips in circles like rodeo lassos, stinging my cheeks, getting into my mouth when I talk. Carven heads toward a beat-up diesel VW Rabbit with the vanity plate "VEGPWR." Popping the hatchback, he reveals his handiwork: in the tire well sits an aluminum fuel tank. He fills it from his jug, sending up a faint smell of Chinese restaurant.

His car is one of several hundred across the country that has been re-jiggered to run on used cooking oil. In the past two years, the greasecar subculture has sprung up fast as a weed. People are doing it as a radical form of recycling. They're doing it to promote fuels derived from vegetable oil, which produce far fewer greenhouse gases than petroleum. They're doing it as a way to protest Big Oil and wars in the Middle East. And they're doing it for kicks.

Grease is the garage band of alternative energy. It's cheap as a beat-up sound system—all you need is a used diesel car and an extra fuel tank. It's loud: most greasecar drivers plaster their bumpers with stickers that proclaim, "Powered by Veggie Oil." And it inspires cross-country trips and lots of couch-surfing, because most greasecar drivers feel compelled to show off the awesomeness of their technology, which can turn trashed vegetable oil into 10,000 miles on the odometer.

Now, Justin and I huddle in the cracked seats of his VW, waiting for heat. He flicks a switch to change fuel tanks, and then we pull out onto the road. We bump along the streets of Florence, Massachusetts, with its Miss Florence Diner and rickety industrial buildings. The VW Rabbit roars, as ancient Rabbits will. Justin has to yell to answer one of my questions. I'm not listening. My mind is choking, sputtering, as I try to convince myself that, yes, we really are being propelled along by the vegetable oil from that jug. We pass a gas station, where salt-smeared cars wait in line for the pump, a sign bearing the price-per-gallon numbers hangs like a prophetic message in the sky. Our gas was free—in more ways than one.

For a second, I flash back to the car that taught me to drive, a gas guzzler with a Deep Purple tape jammed into the 8-track player and three-on-the-tree gearshift. I remember the first time I pulled it onto the highway alone, my hair whipping into my mouth as I watched familiar landmarks fly away. Back then, I wanted to flee from my family. Now, in this VW Rabbit, I feel as if we're escaping from something much, much bigger.

***

Just past the town of Amherst, Massachusetts, you hang a left onto a narrow road and wind through meadows and apple trees.

Soon you'll find yourself among low-slung cardboard-colored buildings that command a sweeping view of sky and snow. This is Hampshire College, known for its excellent drugs. Here, in the fall of 1999, students wandered into a science classroom and found what looked like the red-painted bones of a dinosaur strewn all over the floor, a disassembled tractor. Their assignment: to rebuild the tractor so that it could rely on green energy. Carven, then a senior at Hampshire, signed up.

One day a Vermont inventor pulled into the school's parking lot, his VW Rabbit idling as it burned oil derived from the jatropha nut, which grows in the tropics. As it turned out, West Africans had been running their diesel generators on nut oil for years. Standard wisdom says that you have to prepare oil for the engine by treating it with lye and methanol. But Carl Bielenberg, who sat at the wheel of the VW that day, didn't bother with the flesh-burning chemicals. While working in Africa, he'd found a way to pump untreated oil straight into his engine.

"I could smell the vegetable oil, and the implications were incredible," Carven said. Since then, Carven admits, he has thought of little else besides using "straight" vegetable oil—the kind you can eat—as a fuel.

***

If you store your olive oil in the refrigerator, then you can appreci-ate the problem. Untreated vegetable oil turns solid in the cold, and even at room temperature it can be pretty gooey. So how do you get that ooze to move through the fuel lines? How do you spray it through a nozzle into the combustion chamber without gumming up the works?

The trick is to install a second fuel tank in your car and a switch on the dashboard. That way, you can start the engine up on conventional diesel while you wait for the veggie oil to heat to about 160 degrees. Once you're sure the veggie oil is hot enough, you flip the switch to change fuel tanks. As long as it's liquefied by heat (and well-filtered), the grease will not damage your engine.

"I became obsessed," Carven says. While his classmates drifted off to play Ultimate, he holed up in the lab, hacking engines late into the night. By the end of the year he and his cohorts had trans-formed a van so that it would burn straight vegetable oil. In the summer of 2000, Carven and a buddy took to the highway on a grease-quest, Dumpster-diving behind restaurants for used oil. "You read about Lewis and Clark, and realize it's hard to have that kind of adventure these days, living off the land as you go," he says.

Carven was promoting not just bio-fuels but also a new way to discover America.

"We stopped in Utica to grease up," reads an entry from his friend's Web diary. "McDonalds's warned us their grease was too nasty to use (too many bits of burgers and fries). Wendy's was out of grease, so we hit a local place, Lotta Burger. They said we could have all we wanted but as we were checking it out the cook came out and warned us that there may be lots of water in it so we said, 'bag this,' and hopped over to see the Burger King. We scored 5 gallons."

After a transcendent summer of Amory-Lovens-meets-Jack-Keroauc adventure, Carven decided to make grease his business. Two years ago, he set up shop, selling the fuel-tank-and-toggle kits. He estimates that he's served about a hundred customers. "People call me and say, 'I'm leaving on a cross-country trip in one week and I need a kit.'" He goes all out to comply.

Once, two groups pulled into his parking lot, both demanding service, ASAP. Group 1 had decided that they needed to be in the High Sierra music festival by Friday in a truck they bought yesterday. Group 2: The Liberty Cabbage Theater Troupe, touring with a play about genetically modified food. "I stayed up half the night,"

Carven says, in order to convert both vehicles to grease and get them on the road.

Now, he's swiveling in a stained desk chair in his warehouse office, showing me the website that belongs to one of his competitors. They sell cheapo versions of the greasecar kit. "That will melt," he says, pointing to the plastic tank pictured on the screen. Carven wants you to know that his kits may be more expensive, but there's no way—no possible way—he could cut costs without cutting quality. I believe him. It's 40 degrees in his office and my feet have gone numb. Carven suffers from a wracking cough, brought on by long hours of soldering aluminum in meat-locker temperatures.

"I almost forgot to show you this," he says, handing me an issue of *People* magazine, folded to Daryl Hannah. The actress has just turned her SUV into a greasecar. "Who sold her the kit?" Carven agonizes, staring at the glossy page. "You'd think Daryl Hannah, of all people, would have been able to afford one of mine."

(a version of this story appeared in *Details* magazine)

THE ALTERNATIVE ALTERNATIVE FUEL

Three years ago, Justin Soares stood in the kitchen of the group house he lived in, consulting a recipe as he measured out methanol (a.k.a. wood alcohol), Red Devil brand lye, and some fry grease he'd begged off a local restaurant. He poured the ingredients into a blender and punched "puree." Later, he took the blender out to his driveway and tipped its contents into the tank of his 1981 Volkswagen pickup. Soares, then a student at Oregon State University, had just made his own fuel.

Eventually, he moved his operation to the backyard—partly out of consideration for his seven housemates, who assumed he had been making soap. As his batches got bigger, he began sharing the fuel, called biodiesel, with friends. "I got them hooked," Soares says. In September 2001, he and his friends started a fuel-making co-op called Grease Works, one of perhaps a dozen such groups that have formed around the country in the last few years. To join, you have to own a vehicle with a diesel engine—most likely a VW or a Mercedes—because biodiesel does not work in gasoline engines.

By the following year, the group decided to buy commercially produced biodiesel in bulk. "In the beginning, it might seem romantic to make your own fuel, but pretty soon you realize it's greasy and grimy work," Soares says. Ready-made biodiesel costs about a dollar more per gallon than gasoline, but advocates argue that this is a small price to pay. People who drive around with "No Blood for Oil" bumper stickers feel like hypocrites whenever they gas up at the local Shell station. For them, veggie fuel represents an end to cognitive dissonance. If some of our fuel was grown in Iowa rather than imported, they argue, America might pursue a different kind of foreign policy in the Middle East. And, of course, burning vegetables creates far fewer greenhouse gases than does petroleum.

Taking a principled drive, though, comes with drawbacks. For starters, though the exhaust smells like popcorn, it's not entirely clean. "If you use biodiesel instead of petroleum, you lower almost all the criteria pollutants coming out of your tailpipe," says Shari Friedman, an environmental consultant based in Washington. "But you are increasing nitrogen oxides marginally." In addition, biodiesel is a fair-weather fuel. On warm days, B100—100 percent biodiesel—works fine. But in the cold, most drivers opt for B20, which is mixed with conventional diesel to prevent congealing. Another problem is that less than 1 percent of Americans drive cars with diesel engines.

At least biodiesel works with existing gas-station equipment and cheap old cars. If the U.S. ever did manage to switch to a scheme in which we use hydrogen to power our cars, we'll have to replace every scrap of equipment, from vehicles to pumping stations. But biodiesel is here now, a do-it-yourself dream. Friedman says she just bought an ancient Mercedes that she plans to run on the fuel. "It's something you can do completely on your own, without waiting for the government or the car companies to catch up."

In Germany, where diesel engines power close to 40 percent of passenger cars, more than 1,000 gas stations offer biodiesel at the pump—at a competitive price, thanks to huge tax breaks and subsidies for alternative fuels. But that's Germany. Such generous subsidies are unlikely in the near term in the United States, and that will limit biodiesel's appeal. Americans tend to view higher gas prices as an assault on basic human rights. When Professor Orlando Patterson of Harvard asked 1,500 Americans to define "freedom," most of them talked about the freedom to travel, and many of them mentioned cars. Far fewer mentioned the right to vote.

"You feel very independent," says Friedman, about her biodiesel-powered Mercedes. She uses fuel that her friends made in their backyard—and it's free.

(a version of this story appeared in the *New York Times Magazine*)

First Person:
Stories From My Own Life

Boston Marriage

Liz is explaining the situation to some guy in customer service. "My roommate and I need to network our computers together," she's saying, seated at the other desk in the office that we share.

The word "roommate" jumps out at me. It's an inadequate word, but it's all we have. What else do you call two friends who are shacked up together in a decaying Victorian, run several businesses and one nonprofit group out of its rooms, host political meetings under oil portraits of Puritan and Jewish ancestors, cook kale and tofu meals for all who stop by, go to parties as a couple, and spend holidays with each other's families? If we were lesbians— as people sometimes assume us to be—we would fit more neatly into a box. But we're straight.

In the year and a half we've lived together, I have struggled with the namelessness of our situation. The word "roommate" conjures up a college dorm, scuff marks on the floors from hundreds of anonymous occupants, locks on all the doors, the refrigerator Balkanized into zones where you can or cannot put your food, Death Metal blasting from the speakers down the hall. It means transience and 20 years old. It does not mean love or family.

Words offer shelter. They help love stay. I wish for a word that two friends could live inside, like a shingled house with faded

Persian rugs. Sometimes, in an attempt to make our relationship sound more valid, I tell people Liz and I are in a "Boston marriage." The usual response is, "You're in a what?"

It's an antique phrase, dating back to the 1800s. In Victorian times, women who wanted to maintain their independence and freedom opted out of marriage and often paired up to live together, acting as each other's "wives" and "helpmeets." Henry James's 1886 novel about such a liaison, *The Bostonians,* may have been the inspiration for the term, or perhaps it was the most glamorous female couples who made their homes in Boston, including Sarah Orne Jewett, a novelist, and her "wife" Annie Adams Fields, also a writer.

Were they gay? Was the "Boston marriage" simply a code word for lesbian love? Historian Lillian Faderman says this is impossible to determine, because 19th-century women who kept diaries drew curtains over their bedroom windows. They did not bother to mention whether their ecstatic friendship spilled over into—as Faderman so romantically puts it—"genital sex." And ladies, especially well-to-do ones who poured tea with their pinkies raised, were presumed to have no sex drive at all. Women could share a bed, nuzzle in public, and make eyes at each other, and these cooings were considered to be as innocent as schoolgirl crushes.

So, at least in theory, the Boston marriage indicated a platonic, albeit nerdy relationship. With ink-stained fingers, the Victorian roommate-friends would smear jam on thick slices of bread and then lounge across from each other in bohemian-shabby leather armchairs to discuss a novel-in-progress or a political speech they'd just drafted. Their brains beat as passionately as their hearts. The arrangement often became less a marriage than a commune of two, complete with a political agenda and lesson plan.

"We will work at [learning German] together—we will study everything," proposes Olive, a character in *The Bostonians,* to her ladylove. Olive imagines them enjoying "still winter evenings under the lamp, with falling snow outside, and tea on a little table, and successful renderings…of Goethe, almost the only foreign author she cared about; for she hated the writing of the French, in spite of the importance they have given to women." James poked fun at Olive's bookworm passion. But he lavished praise on his own sister Alice's intense and committed friendship with another woman, which he considered to be pure, a perfect devotion.

Most likely, the Boston marriage was many things to many women: business partnership, artistic collaboration, lesbian romance. And sometimes it was a friendship nurtured with all the care that we usually squander on our mates—a friendship as it could be if we made it the center of our lives.

"I am on my way through the green lane to meet you, and my heart goes scampering so, that I have much ado to bring it back again, and learn it to be patient, till that dear Susie comes," Emily Dickinson wrote to her friend—and maybe lover—Sue Gilbert. Today I see tragedy in these words, for Sue ended up married to Emily's brother, and the women never had a chance to build a life around their love. I find myself wishing I could teleport them to our own time, so that Emily D. and her Susie might find an apartment in San Francisco together, fly a rainbow flag out front, shop at Good Vibrations, and delight one another with dildos in shocking shades of pink. And yet, it's not that simple. When I read the passionate letters between nineteenth-century women, I become keenly aware of what I'm missing, of how much richer Victorian friendships must have been. While our sex lives have ballooned in the last hundred years, our friendships have grown stunted. Why don't I shower my favorite girls with kisses and "mash" notes, hold hands with them as we skip down the street, or share a sleeping bag? We don't touch anymore. We don't dare admit how our hearts scamper.

Several years ago, I fell in love with a man because of all he carried—he would show up for the night with five plastic bags rattling on his arm, and then proceed to unpack, strewing possessions everywhere. The next day, I'd find his orange juice in the refrigerator, his sweater tucked into my bureau, a software program installed on my computer. Night after night, he installed himself in my apartment.

At first, every one of these discoveries charmed me—his way of saying, "I need to be with you." But one morning, I surveyed my bedroom—guy's underwear on the floor, books about artificial intelligence stacked on the night table, a jar of protein powder on the shelf—and realized that I had a live-in boyfriend. And that he and I had completely different ideas about what we wanted from a living space. He thought of an apartment as a desktop where we could scatter papers, coffee mugs and computer parts. What I regarded as a mess, he saw as a filing system that should under no circumstances be disturbed. Meanwhile, I drove him crazy by hosting political meetings in our living room, inviting ten people over for dinner at the last minute. We loved one another, but that didn't mean we should share an apartment.

And then—when our Felix-Oscar dynamic seemed insurmountable—I picked up a magazine called *Maxine* and stumbled across an article that gripped me. Written by 27-year-old Zoe Zolbrod, it celebrated the passion that flashes up between women, even when they are both straight: "I would meet women who I would need to know with an urgency so crushing it gave the crush its name. And in knowing them I would feel a rush of power and possibility, of total self, that seemed much more real to me than heterolove," Zolbrod wrote. When she met her friend V, "it was like finding the person you think you'll marry." The two moved in together. They took care of each other, became family, called each other "my love" and "my roommate" interchangeably.

I remember reading that article and thinking, "yes." I adored my boyfriend, but he and I had never meshed in the way that Zolbrod described. We tried to make a home together, but we didn't agree on what a home should be.

Years later, when our love fizzled into friendship and he moved out, I made a vow to myself: I would not drift into a domestic situation again. Instead, I would find someone who shared my passion for turning a house into a community center—with expansive meals, weekend guests, clean counters, flowers, art projects, activist gatherings, a backyard garden, and a pile of old bikes on the porch, available to anyone needing to borrow some wheels.

My friend Liz seemed like the right person. And so I proposed to her. Did she want to be a co-creator of the performance art piece that we would call "home"? She did.

Recently, at a party, I met a thirtysomething academic who has settled alone in a small town outside of Boston. "I can step right out my door and cross-country ski," she told me. "But I'm lonely a lot." Around us, people sweated and threw their arms wildly in time to an old Prince song. The academic wedged her hands into her jeans pockets, and her eyes skated past my face and scanned the room.

If you're lonely, get a roommate, I suggested. Move into a group house. "No," she sighed. "I'm too old for that. I'm set in my ways." What if you marry? I asked. She laughed. "That's different."

She might be speaking for thousands, millions of women all over this country. According to the U.S. Census Bureau, one out of four households in 1995 had only one member, a figure expected to rise sharply as the population ages. I see the future of single women, and frankly, it depresses the hell out of me. We're isolating ourselves in condos and studio apartments. And why? Sometimes because we need to bask in solitude—and that's fine. But other times, it's because we're afraid to get too comfortable with our friends. What if you bought a house with your best friend, opened a joint bank account with her, raised a child? Where would your bedmate fit into the scheme? This is where the platonic marriage—for all its loveliness—may force you to make some difficult choices and rethink your ideas about commitment.

Liz's love, a theoretical physicist, meanders down our street clapping. Standing beside a triple-decker house, he cocks his head, listening to the sharp sounds reverberating off of a vinyl-sided wall. He's designing an exercise for the students in the "Physics of Music" class that he's assistant teaching. When he's done, he'll come back inside to find Liz and me draped across the sofa, discussing urban sprawl. We'll all make dinner together, and if I feel like it, I might join them for a night out, or I might head off with the guy that I'm seeing.

I date scientists too, men who understand what it is to experiment, to question and wonder. Liz's love or mine might sit in our kitchen scrawling equations into a notebook, or disappear for days to orbit with subatomic particles or speak with machines. These men are wise enough to see that the Boston marriage works to their advantage. Liz and I keep each other company. Our Boston marriage has made it easier for us to enjoy the men in our lives.

But how do we commit to each other, knowing that someday one of us may marry? One of us might fall in love with something other than a man—a solar cabin in Mexico, a job in Tangier, a documentary film project in Florida, a year of silence in the Berkshire woods. Any number of things could pull us apart. We have made no promises to each other, signed no agreements to commit. For some reason, that seems OK most of the time.

For this article, I talked to many women who'd formed platonic marriages or who'd thought about it seriously. All of them discussed the complicated issues of commitment, or lack thereof, between friends.

Janet calls her arrangement with Greta intentional. "In the same vein as creating an 'intentional community,' we have an 'intentional' living arrangement," she says. The two high school friends, both straight women in their early thirties, moved to Boston together five years ago, knowing that they would share an apartment, and a life. They eat dinner together and check in with the how-was-your-day conversation most people expect from a mate.

"Greta is the person I say to contact when I fill out emergency cards," Janet tells me. "She is the first person I would turn to if I needed help. "

And yet, the two have left their future open, and the promises they have made to each other are full of what-ifs. If Greta doesn't marry by the time she's 35, they might raise a child together. It's the what-ifs that drive many women away from closeness with each other.

One married woman, I'll call her Lisa, says she's deeply disappointed with the way women treat their friendships as disposable, dumping friends when an erotic partner comes along. "Even though my friends and I used to talk about buying a house together, we all knew at some level that it wasn't going to work. Ultimately, we would betray each other, find a man, marry him. I got married because I knew everybody else was going to. If I knew I could trust a friendship with a woman—that there was a way of making a friendship into a bona fide, future-oriented relationship—I would rather have that than be married."

As for me, I've come to think of commitment as something beyond a marriage contract, a joint bank account, or even a shared child. I know that eventually Liz and I may drift to other houses, other cities. Yet I can picture us reuniting at age 80, to settle down in an old-age home together. Maybe we will have husbands, maybe not, but we'll still be conspirators. We'll probably harangue the youngsters who spoon spinach onto our plates about the importance of forming a union; we'll attend protests with signs duct-taped to our walkers; maybe we'll write an opera and perform it using some newfangled technology that lets us float in the air. Liz and I are committed. We share a vision of the kind of people we want to be and the world we want to inhabit.

"We formed a family core with the possibility of exhilaration," wrote Zoe Zolbrod in her article. "Yet Hallmark never even named a goddamn holiday after us, can you believe it?" We're not sure what to call ourselves. We have no holidays. We don't know what our future holds. We have only love and the story we are making up together.

Liz sashays into the kitchen, a shopping bag crinkling under her arm. "I bought you these," she says, "because you've been wearing those mismatched gloves with holes in them."

I slide on the mittens, and my hands turn into fuzzy paws, pink and red with a touch of gold. "I love them," I say, and hug her, patting her back with my fuzz. She laughs and shifts her eyes away, a bit embarrassed by her own generosity. "I couldn't have my roommate going around in shabby gloves," she says.

She uses the word "roommate." But I know what she means.

UPDATE:

Liz and I liked to believe that our utopia would go on indefinitely, but instead it lasted four years, which is still a good long time. In 2003, her boyfriend took a postdoc in Manhattan and she left to join him in a fabulous apartment in The Village. By that time, I was spending most nights over at my boyfriend Kevin's house anyway. So Liz and I each ended up ensconced in hetero-normative relationships.

In 2004, Massachusetts legalized same-sex marriage. Days after the ruling, my friends Julie and Laura took vows under a fluttering hoopa, and a rabbi announced that their partnership was now protected by the laws of the Commonwealth. I wept with joy -- as did almost everyone sitting in the folding chairs around me.

Liz and I remain close friends, and we still call each other "my Boston wife."

The Encyclopedia of Scorpions

"Are those polka dots?" I called, trying not to sound like I was panicking.

Nancy, in the front of our kayak, twisted to peer back at me. "What?"

I pointed ahead. We were floating toward a stretch of sand that was pocked with blotches, as if a giant clown had laid out his suit to dry.

"Weird," Nancy said, completely unconcerned about the dots on the beach. She stowed her paddle and—as if she were dismounting from a circus pony—threw herself out of the kayak and into the water. She began pulling the boat toward shore, staggering through the waist-high surf, her legs sending up arches of glitter, her tiny safety vest riding high on her shoulders.

Then, with the sea lapping around our knees, we waited for the others. We couldn't carry our kayak onto the shore by ourselves—it was laden down like a suburban station wagon, avocados and bananas and dry bags and sacks of water strapped to its every surface. The waves threw pebbles onto the shore and then pulled them back again, with a sound like breaking glass. I shaded my eyes and studied the beach. The splotches appeared to be rocks, but they came in colors that alarmed me: the taupe shade of a Band-Aid, the purple of a bruise, the blush of fevered skin.

Soon, the rest of the kayaks floated up, the boats nuzzling against each other, people splashing into the water. Sam, our guide, untied a rope from the back of his kayak and pulled what looked like a boombox out of the water: our latrine. He had forbidden us from defecating in the desert because of a profound dryness that would harden our turds into joke doggie poop, creating an eco-hazard. When we left a campsite, Sam towed the boombox behind his kayak and emptied our shit out at deep sea. This service alone, it seemed to me, was worth the price of the trip.

***

Nancy and I had known each other for years. We belonged to the same tribe of friends, people in their 30s who took vacations together, finding some cabin in the mountains where we could hold a grownup slumber party, with tequila and poker. Always, Nancy and I would grow restless in the cabin. We'd wake up early, the others still breathing softly in their sleeping bags, and slip out with our cross-country skis, reappear five hours later with the tips of our hair frozen stiff.

A few months ago, I'd called her. It was one of those February days when the salt from the last snowstorm smears the asphalt and darkness comes at four. "Do you want to fly out West and go hiking?" I said, the cordless phone wedged into my neck. This was my craving: hours climbing up a sandstone mountain with orange dust stuck to my skin, then a beer, some Mexican food.

"You know what we should do?" she said. "We should take a week and go down a river in kayaks."

I realized that she, the former leader of Outward Bound trips, envisioned us navigating by the stars, paddling alone on the open sea, stopping on windswept shores to forage for berries. She would not agree to any version of camping that involved watching cable TV at night.

So we compromised: we'd sea kayak, but with a guide and a group. I figured nothing could go too wrong.

***

Now, Nancy and I were trooping down the beach with infected-looking stones, our dry bags bumping against our legs. We were off to set up our tents for the night, away from the others. The social dynamics of our kayaking group reminded me of Gilligan's Island.

Remember how the castaways never quite formed one unit, but instead clung to the identities they'd had before the wreck? Skipper, movie star, millionaire.

In our castaway society, Nancy and I were the Superwomen, and we maintained a polite distance from the Biologists, the Adman, and the New Age Divorcee. We'd become the Superwomen because our kayak always shot out ahead of the group, and the guide had to yell at us to turn around and come back where he could see us. Nancy, of course, supplied 90 percent of the muscle.

She dumped her stuff on the sand. "Hey," she said, as she bent over, "can you name all the American presidents, in order?"

"Washington, Jefferson, Madison," I began, unpacking my tent. By the time I'd reached Polk, Nancy had hers set up—white fabric dazzling in the sun. It rippled a bit in the wind, but was otherwise as tight as a sail.

I was still arranging a mess of orange and purple nylon on the sand. I'd borrowed the tent from an ex-boyfriend; he'd neglected to warn me that this impressive piece of high-tech gear was designed for ice. It could wedge into the crevice of a mountain, its stakes hammered into frozen snow. A clear-plastic window trapped heat from the sun. A vent made out of fabric—essentially a shirt-sleeve that extended from the side of the tent—allowed you to fire up a stove inside, without asphyxiating on its fumes.

I coaxed the tent onto its knees. It was a rumply, wrinkled mess, because its stakes did not hold in the sand. When I crawled inside it to change out of my wet clothes, the sun clawed through the clear-plastic and sweat immediately popped out on my skin. It had to be 105 in there.

Sam had lectured us about keeping our tents closed up at all times. Otherwise, he said, "varmints" would get in. But after two days without any varmint sightings, we'd all grown cavalier. Now, I opened up the vent to a deliciously cool circle of blue. I unzipped the tent door to get a breeze moving through. I left the tent that way, and strolled off.

***

An hour later, I was walking out of the ocean backwards with a snorkeling mask on my face, my flippers thwacking the wet sand. One of the Biologists called out to me, but the wind tore his words away. I could hear, though, the panic in his voice.

"What?" I called.

"Nancy," I heard and "scorpion."

I shucked off the flippers and ran up the beach, my cheeks still throbbing where the snorkeling mask had pressed. In the sand, my sprinting gait turned into the slow-motion plod of nightmares where you can never run fast enough.

She sat on a puke-colored rock, holding one hand up against her chest, looking even tinier than usual, those shorts that would fit a twelve-year-old girl, those sinewy legs extended, her head big on her body, curls beating at her cheeks. She must have been clenching her teeth—her mouth looked all wrong.

"There was a bug on my shirt. I brushed it off. Can you believe I did anything so stupid?" She shook her head.

Sam, the guide, arrived with a leather-bound book, his finger marking a page. At first I took it for an encyclopedia, but then he said, "see if you can identify it by the picture," and held the book in front of her. Over her shoulder, I saw the engravings. This was not an encyclopedia of words. It was an encyclopedia of scorpions. And also of rattlesnakes, of heart attacks where there are no defibrillators, of blood poisoning and parasites and broken bones. The book seemed to be smeared with all the bad news it carried.

Nancy examined the engravings and finally pointed with her good hand.

"You sure?" he said, poker-faced.

"Pretty."

"Unfortunately, that's the worst of the lot," he said. "It sometimes kills kids and old people, but you're going to be OK. If you were going to have a reaction, you probably would have had it by now."

The beach, dimming now, the sand turning from white to gray, seemed to warp, to pulse along with my heartbeat. Sam said the scorpion wouldn't kill a healthy adult. But Nancy was so small. She weighed no more than a hundred pounds.

"We've got to get her out," I told him.

"We can't," Sam said, closing the book and carefully sliding it into a dry bag.

"Of course we can. There must be people somewhere."

Sam shook his head. "There's no way to evac." He pointed to a shrimp boat parked out near the horizon, its nets folded like bat wings. "The best we could do is try to radio that boat, but there's probably no radio on board."

"I'm OK," Nancy said, and she slid off the rock and onto the sand, rubbing some of it over her bad thumb, as if this were medicinal. "It feels like I stuck my finger in a light socket. But I can stand it. It's tolerable."

Something about her calm unsettled me all the more.

"Don't you have anything to give her?" I demanded.

He squatted beside her. "Look, she'll be OK. The Mexicans get stung all the time and they don't take anything for it."

In other circumstances, I would have sided with him— denouncing spoiled Americans and the way they ran whining to emergency rooms and demanded painkillers. But panic brought out the first-worlder in me. I believed that just because we were Americans, a hospital should materialize in the desert, its sliding doors swooshing open to frosty air-conditioned air, end tables piled with *People* magazines, doctors padding around in their surgical booties. Nothing materialized. Instead, the beach darkened. The waves made their death rattle.

"What's in your first-aid kid?" I tried, "An epi pen? Antihistamines?"

He ignored me. "Do you think you can make it without aspirin?" he asked Nancy. "It would be better to let your body clean out the poison, without adding anything else into the mix."

"Yeah. This is the strangest feeling. Buzzing. Like there are bees inside. And the weirdest part is, it's slowly going up my arm."

He nodded. "It's moving to your heart."

"It's here now," she pointed just above her elbow.

He continued to nod as she spoke, and I guessed that rather than paying attention to her words he was listening for slurred speech.

She began to ask him something, but he interrupted.

"Nancy, let's get you out of that sweater," he said. And then, like a gentleman helping his date take off her wrap, he maneuvered her Polartec sweater backwards off her shoulders.

I assumed that he wanted to examine the arm. But instead, he snatched the sweater and danced a few feet away from us. He trained his flashlight on the fabric as he shook it. Something small and yellow fell off. He followed the dot with his flashlight, but it had disappeared into the sand. "Another one," he said. "They're everywhere. They live under those rocks." He pointed with his light at one of the strange stones. "Now that it's night, they're going to come out."

"I guess I shouldn't be sitting here," Nancy said, struggling to her feet.

We ate dinner huddled on a blue tarp, all in a clump, nobody so much as sticking a toe out onto the sand. Except Sam. He squatted in front of the fire, putting a pot of water on the grate, so we could wash dishes. His face, flickering yellow, was impassive as the cliffs that we'd paddled past today. Was he really sure that Nancy would be OK? Maybe he was only pretending, so as not to spread fear.

"Where's it up to now?" I asked Nancy, who sat crunched beside me on the tarp.

"Here," she said, touching just below her shoulder.

"What about when it gets to your heart?"

"It'll be OK." Her voice floated up to me through the dark. "It's starting to throb less."

"You didn't eat anything."

Instead of answering, she lay down onto the tarp. The Adman readjusted his legs to make room for her. "I just want to look at the stars right now," she said.

Later, I walked her back to her tent. We switched off our flashlights and listened to the surf for a while.

"You look miserable," I said, though I couldn't see her face. The only thing visible besides the salted sky was the winking of the campfire, way down the beach. I thought I heard scuttling in the sand all around us, those tiny demons in their armor, dragging stingers behind them. I hated this place.

"Really, I'm fine," Nancy insisted. "My whole arm is buzzing like it's fallen asleep times a hundred, but I wouldn't say it hurts." She paused for a moment, and I heard her shift something. "I've never been stung by a scorpion before," she said. "It's kind of cool."

"Maybe I should sleep in your tent tonight, just in case." In one corner of my brain, I knew Nancy would be OK by morning; in another corner, I was convinced she might die in her sleep.

"I'm all right," she said.

"You sure?" I wanted to stay up all night beside her, with my flashlight trained on her arm, to watch for any new symptoms. Even better would be if I could wake her up every hour or so and ask her whether she felt better. But I knew she would never allow that.

"I'll yell if I need anything." She leaned down and I heard her unzip her tent.

I flicked on my flashlight; in the profound dark, it illuminated only a small circle of sand, the pebbles and twigs cast their own mini-shadows. I was exhausted. I stumbled along, training the flashlight this way and that, until the tent appeared in the spotlight. It looked like an old drunk sleeping on a sidewalk grate, half-collapsed, a mess of wrinkles and folds. The open vent waved at me. The door flapped, yawning wide.

I remembered now: I had left everything open. Scorpions loved to crawl inside and under. An army of them could have filed through the gaping holes in my tent. I would have to examine every last inch inside.

I duckwalked through the door—I didn't dare put my hands or knees down in the gloom—and shone the flashlight at my dry bags, books, sleeping bag, t-shirt. I was so worn out that I could have happily collapsed into my pile of belongings. But I wouldn't allow that. The sleeping bag, for instance. I should take it outside and shake it, then unzip it and examine every fold. I picked it by pinching its nylon shell and tried to steer it outside. The bag bunched up, got caught on a zipper. Forget it, I thought, and heaved the sleeping bag into the dark. It landed somewhere invisible with a sad rustling sound. I figured I'd deal with it in the morning. I picked up the books and tossed them out too. Next went my clothes. And the extra pair of shoes.

Then, with all my possessions gone, I squat-walked around the edges of the tent, checking under folds and up the walls, in every crevice and wrinkle. Finally, when I was reasonably sure that I'd secured a scorpion-free environment, I lay down. I could feel the grains of sand shift under the nylon. Without a pillow, my cheek flattened against the ground. My neck ached. I shivered. I pulled the hood of my sweatshirt over my head. The starlight seeped through the plastic window. The tent made polite rustling sounds, as if it were trying to adjust itself without waking me.

That night, I didn't have dreams so much as revelations. At one point, I heard a faint gurgling underneath the sand, a slurpy intestinal sound. It was then that I understood the purpose of the oddly colored rocks on the beach. They blocked the entrances to the bowels of the earth. And in a flash, I knew the terrible and most essential fact of life could be summed up in one word: digestion. Right now, in the desert, insects crawled over a coyote carcass and gnawed it clean. Right now, inside my skin, cells shredded and sorted. The tent itself was a digestive apparatus, an intestine of sorts, that was melting me down into a nub. My tent was sucking me away, like a candy. Soon I would be gone.

I startled awake. I was back in the real tent, back in my ordinary sense of my body, with its damp feet and achy back.

A relief, except that I was still on a hellish, vermin-infested beach. I had chosen to be here. That was the crazy thing. I was the one who had done this to me.

***

When I woke, I knew by the angle of the sun that it must be past 7 in the morning. Why hadn't anyone come by to wake me up? I could hear nothing but the thrashing of the surf. I scooted out of the tent. My possessions lay on the white blankness of sand as if they'd been posed there by a photographer for some surreal art-rock album cover, the sleeping bag curled in a C, a book face down, a flipper. Nancy's tent had disappeared. It was as if she had never been there.

Way down the beach, people milled about, clumping together and then coming apart. The sea flashed with light, stabbing my eyes. The people, silhouetted against the waves, seemed to wink in and out of existence. I thought I saw one of them quivering with sobs. Out past the rolling surf, Sam's red kayak bobbed, heading toward the horizon. Something had happened. I sprinted down the beach, sand flying up and sparking against my calves.

But then, I slowed to a trot and snorted with relief. As usual, I'd been too quick to panic. They were packing. One of the Biologists stood over the pile of water sacks, his finger pecking the air, figuring out how many we'd each have to carry. Nearby, Sam dragged the latrine along behind him, like a pull-toy. And the red kayak out in the cove, bobbing in the waves? That was Nancy. I knew her by her hair. It flew around her face in a whirlwind.

"Hey," I yelled, waving my arms. And then, because she hadn't seen me yet, I jumped up and down. "Hey, hey, hey." And every time I landed, the sand caught me.

"What are you doing?" Sam scolded. "We're just about ready to go."

So I ran up the beach again, to my fallen tent. I picked up a book, shook it vigorously, and sand rustled from its pages, but no scorpion, no scorpions anywhere, for it was a new day and I had just chugged the last of the coffee from a plaid thermos like the ones in 1950s *Life* magazines, and we were not dying in the desert, we were on a lark! Now I held snorkel mask in my hand—it reflected the sky, the blue leaping across its surface. It sent up a perfume of the ocean, the briny smell of childhood vacations. A cactus in the distance caught sunlight in its spines, glowed with a golden nimbus. Just when I had no time for it, everything around me had swelled into beauty. I picked up the sleeping bag, and remembered how it terrified me the night before, with all its folds and secret places. Now it was warm as dough. I shook it out gaily, enjoying the way the breeze took over and lofted it up. Then I stuffed it into its sack, without worrying too much about my hands disappearing into wads of fabric.

I peeled off my sweatpants and took possession of the air with my bare legs. I was out in the middle of the world in my underpants, a t-shirt and sneakers. For a moment, I flirted with the cliff beside me, and I let the ocean see a little more of my ass than was proper. Then I picked up a pair of shorts, gave them a cursory shake and lifted one leg to step into them.

I froze. I slowly lowered my leg, and adjusted the shorts ever so carefully so I could peer down inside. A scorpion clung to the seam just below the waistband. It was the gold color of sand, half the length of my pinkie. Its tail curled so tightly that the creature looked like it would sting its own head. My heart began to pound everywhere in my body—my cheeks, my feet, my eyes. I had come so close to putting my leg into that tunnel of fabric.

I shook the shorts gingerly. Nothing fell out. I tried again. In the end, I had to pull on my sweatpants and go fetch Sam. He followed me with a stiff-legged gait of annoyance, picked up the shorts, sized up the scorpion, and gave a shake. I expected it to fall out immediately, cowed by him. But it didn't. He flapped more vigorously. Then he found a stick and used it to fling the shorts back and forth in front of him. The shorts flew at waist height, whooshing up specks of sand that turned into glitter in the air. Then he waved it up high, against the blue, so that I was afraid the scorpion might fall on his head. Sam seemed to be intoxicated, as if he were waving a flag after a revolution, the flag of his own newly formed country. It would be a land of gringos with their hair burned blonde, and wilderness guides who carry tidal charts in their shirt pockets, and people who figure everything will be OK once the poison reaches their heart. This was Nancy's country too. But not mine. My country, for better or worse, was America.

Sam checked inside the shorts again and sighed, to let me know that the scorpion was still there. "We're going to have to use water," he said, and I followed as he carried the shorts down to the ocean. He balanced on a boulder, leaned out, and dropped them into a shallow place.

I clambered up onto the boulder beside him, feeling its roughness, like super-tough sandpaper, against my hands.

"I hate to do this," he said. "It's wrong to kill them. They have more right to this place than we do."

The fabric of the shorts turned from blue to black and sunk under the surface of the water. A bedraggled scorpion emerged from the folds. Sam thrust the stick in its path, so it could climb to safety. "Here, please," he said to it. On one last island of dry fabric, the scorpion stood its ground, threatening Sam's stick with its whip of tail. It was a tiny clenched fist of a being, all anger and swagger. The island under its feet darkened and sunk and water welled in, and the scorpion melted, as witches are supposed to do, though it didn't entirely disappear. A soft yellow rind remained, bobbing on the surface of the water like a bit of egg, a crumb, a curd, something stuck to the edge of a plate that comes off in the sink. And then what was left of the scorpion tumbled away and it was gone.

Off Season

Our family may not be prettier than yours. We may not be smarter. But we do know how to avoid traffic jams. "You'd have to be a lunatic to get on the Pike after 3 o'clock," my mother used to say to me, when I was too young to understand the word "lunatic." But I took her meaning. There were people out there pushing and shoving to get to the same place at the same time. We would not be among them.

My mother has a theory: eccentricity is efficient. When I was about eight years old, she hoisted a Turkish flag onto the antenna of the family station wagon. "This way, when we park in the mall, I'll always be able to find the car," she explained. "Who else is going to have a Turkish flag?" And so, for several years, my family traveled around like some rogue embassy, the sickle-moon of a faraway nation fluttering above as we drove, motorcade style, on a mission to find a new set of bed sheets.

Years later, I sprawled across a beanbag chair in my high-school library, studying *People* magazine. Eleven kids—about my age— had been smothered at a Who concert. As I examined each of their photos, lined up yearbook-style on the glossy pages, I was gripped by an uncharitable disdain for those kids. I might be an oddball, some might say a geek, but at least I wouldn't have been caught dead in a stadium-rock hall in Ohio, wearing feathered hair and designer jeans. Those kids had in fact been caught dead in just such circumstances. And that seemed like the real tragedy.

At the time, of course, I thought I had developed such opinions on my own. But now it's clear to me that my devotion to the uncrowded and off-season, the underground and the strange, derived from my mother's obsession with traffic. Always, always avoid the crowds.

***

A decade ago, my father lay in a rented hospital bed in my parent's house, dying. My mother and sister and I took care of him, circling around the room fetching Chopin CDs or hand lotion or morphine—anything to mute his pain. It was late November, that stretch of days that dangles between Thanksgiving and Christmas. Even if your father isn't dying, it's easy to feel awful in that ditch between the holidays, with the frayed leaves eddying around your feet. *It's A Wonderful Life* plays round-the-clock on cable, mocking you with repeated visions of Jimmy Stewart cocooned by his brood of devoted children. Psychiatrists' schedules fill up. Everyone is convinced that everyone else is going home to a happy family.

That year, it seemed I was always driving to the grocery store to fetch cans of protein drink, which was all that Dad could keep down. "Rudolph the Red-Nosed Reindeer" echoed down the produce aisles. Signs wished the shoppers happy holidays and joy to the world. The food itself seemed like just more gaudy holiday decoration—the dollops of turkey and the confetti-ed cookies and the cranberry sauce, which would roll out of its can like a giant ruby. Dad would never, ever get to eat any of this stuff again.

He died in early December. While other people were decorating their trees, we shopped for a headstone. After the funeral, we sprawled in the living room too tired to put away the cheese plates. "Girls," my mother said to us, "let's not have Christmas this year." We'd always been fuzzy about holidays, moving around dates to suit our needs, but I had no idea that you could cancel Christmas entirely. My mother did know this. That is her genius. She never seemed so brave to me as at that moment when she—wearing slippers and a sweatshirt, pale, exhausted—decided we didn't have to do what other families did. At that moment, she hoisted a flag over us more exotic and beautiful than the Turkish moon. The three of us, she seemed to say, would become a little country all our own. No apologies. No regrets. No rush hour.

ABOUT THE AUTHOR

Pagan Kennedy is the author of ten
books in a variety of genres—from
cultural history to biography to the
novel. A regular contributor to
the *Boston Globe,* she has published
articles in dozens of magazines and newspapers, including several
sections of the *New York Times.* A biography titled *Black Livingstone*
made the *New York Times* Notable list and earned Massachusetts
Book Award honors. Her most recent novel, *Confessions of a Memory
Eater,* was featured in *Entertainment Weekly* as an "EW pick."
Another novel, *Spinsters,* was short-listed for the Orange Prize. She
also has been the recipient of a Barnes and Noble Discover Award,
a National Endowment for the Arts Fellowship in Fiction, and a
Smithsonian Fellowship for science writing.

www.pagankennedy.net
www.dangerousjoy.com